Multiscale physical processes of fine sediment in an estuary

Multiscale physical processes of fine sediment in an estuary

DISSERTATION

Submitted in fulfillment of the requirements of
the Board for Doctorates of Delft University of Technology
and of
the Academic Board of the UNESCO-IHE Institute for Water Education
for the Degree of DOCTOR
to be defended in public
on Monday, 8 June 2015 at 15.00 hours
in Delft, the Netherlands

by

WAN Yuanyang

Born in Hubei Province, China

Bachelor of Engineering, Wuhan University, Wuhan, China
Master of Science, Changjiang River Scientific Research Institute, Wuhan, China

This dissertation has been approved by the
promotor: Prof. dr. ir. J.A. Roelvink

Composition of the doctoral committee:

Chairman	Rector Magnificus, Delft University of Technology
Vice-Chairman	Rector of UNESCO-IHE
Prof. dr. ir. J.A. Roelvink	UNESCO-IHE/TU Delft, promotor

Independent members:

Prof. dr. ir. Z.B. Wang	Delft University of Technology
Prof. dr. ir. J.C. Winterwerp	Delft University of Technology
Prof. dr. ir. M. Chen	Vrije Universiteit Brussel, Belgium
Dr. A. Sottolichio	University of Bordeaux, France
Prof. dr. John Z. Shi	Shanghai Jiao Tong University, China
Prof. dr. ir. A.E. Mynett	UNESCO-IHE/TU Delft, reserve member

This study was supported by Shanghai Estuarine and Coastal Science Research Center and UNESCO-IHE Institute for Water Education. Funding was provided by UNESCO-IHE Partnership Research Fund (UPaRF, No. 60038881), Shanghai Municipal Natural Science Fund of China under grant No. 11ZR1415800 and National Key Technology R&D Program of China under grant No. 2013BAB12B00.

CRC Press/Balkema is an imprint of the Taylor & Francis Group, an informa business
© 2015, WAN Yuanyang

Published by:
CRC Press/Balkema
PO Box 11320, 2301 EH Leiden, The Netherlands
e-mail: Pub.NL@taylorandfrancis.com
www.crcpress.com – www.taylorandfrancis.com
ISBN 978-1-138-02844-9 (Taylor & Francis Group)

To my family

Summary

Estuaries are natural highly dynamic and rapidly changing systems, comprising a complex combination of physical processes on many different time- and space- scales. Fine sediment physical processes are attracting increasing attention by coastal engineers. One reason is that we need more coastal reclamations for sustainable society development and more and more harbors, ports and navigational channels. These are increasingly constructed in those fine sediment surroundings, where used to be considered as unfavorable place for waterway development, due to the high possibility of confronting with sedimentation issue. Another reason is because we are interested in the pursuit of the fascinating nature of fine sediment dynamics.

The Yangtze Estuary is an excellent example of a fine sediment estuarine system with a moderate tidal range (~1-5 m) as well as a highly seasonally-varying (~10000-50000 m³/s) freshwater inflow. The sedimentation ranks as a key issue in the Yangtze Estuary recently. Before conceiving a measure to mitigate channel siltation, the reasons and mechanisms related to the characteristics of sedimentation should be investigated as the first step. Thus, understanding the underlying mechanisms associated with fine sediment transport, ETM (estuarine turbidity maxima) dynamics and sediment trapping in the Yangtze Estuary is considered as a major challenge to maintain the "golden waterway".

According to the systematic study on the topic of multiscale physical processes of fine sediment in a meso-tidal convergent alluvial estuary, the main contents and conclusions are summarized as follows.

In *Chapter 1*, **the complexity of fine sediment transport in estuaries is briefly identified** Apart from the interactions among riverine inflow, oceanic tide, wind wave, the Coriolis force, saline water intrusion and bed resistance in an estuarine system, the micro-scale effects from flocculation and hindered settling, baroclinic forcing, turbulence damping and drag reduction show the obvious influences on internal structures of current, salinity and suspended sediment concentration (SSC) which in turn have an impact on the macro-scale current and sediment regimes and morphological evolution. After highlighting the unique features and the challenge of the Yangtze Estuary, the main objective and organization of this study are introduced.

Chapter 2 focuses on the recent decadal hydrodynamic evolutions in the whole Yangtze Estuary from a series of hydrological data. Over the past few years, the Yangtze Estuary has witnessed an unprecedented scale of human intervention through extensive resource utilization. We found that, (i) the water level along the main outlet of the Yangtze Estuary increased from 1998 to 2009; this increase was induced by the variation in the whole river regime (including natural morphodynamic processes and local topography feedbacks from extreme meteorological events and human activities); (ii) the decrease of the flow portion ratio at the 3rd bifurcation is directly induced by the Deepwater Navigational Channel (DNC) project and the corresponding morphological changes at the North Passage; and (iii) the estuarine environmental gradients (salinity and suspended sediment concentrations) were compressed, and the fresh-salt gradient became steeper. **This has the indirect effect of back-silting on the waterway, i.e., strengthening the stratification effect near the area of estuarine turbidity maximum and enhancing the tendency of up-estuary sediment transport.**

In *Chapter 3*, observations of storm-induced fluid mud dynamics have been conducted at the DNC of the Yangtze Estuary from October to December 2010, during the occurrence of a cold-air front. The observed data reveal that just after the critical wind wave event, a large amount of fine sediment was trapped in a state of fluid mud along the channel. The observed thickness of the fluid mud was up to about 1-5 m, which caused some significant economic and safety problems for shipping traffic in the Yangtze Delta area. The mechanisms and transport processes of the storm-induced fluid mud are analyzed and presented from the angles of both process-oriented and engineering-oriented methods. With the help of tidal hydrodynamics and wave modeling, it could be inferred that **the behavior of the storm-induced fluid mud event mainly depends on the overall hydrodynamic regimes and the exchanges of sediment, which is released by storm-wave agitation from adjacent tidal flats. These sediments are accumulated as fluid mud, and subsequently oscillate and persist at those locations with weaker longitudinal residuals in the river- and tide-dominated estuary**. In addition, the downslope transport of fluid mud is also thought to have stimulated and worsened the fluid mud event observed in this study. Our modeling results and observations demonstrate that: (i) the transport of fluid mud is an advective phenomenon determining the central position of fluid mud layer along the channel, and it's also a tidal energy influenced phenomenon controlling the erosion and accumulation of fluid mud; and (ii) both suspended particulate matter availability and local residual flow regime are of critical importance in determining the trapping probability of sediment and the occurrence of fluid mud.

In *Chapter 4*, spatial and temporal measurement data describing spring-neap variations of velocity, salinity and SSC in the DNC of the Yangtze Estuary were obtained in the wet season of 2012. These data were collected in the middle of the DNC for the first time, and apparently document the formation of a rather stable density stratification interface and salt wedge, especially during neap tides and slack waters. The convergent zone of residual currents, salinity and sediment during neap and spring tides oscillates in the middle and lower reach of the DNC. It encourages the formation of a near-bed high-SSC layer, which favors siltation in the dredged channel. Both the near-bed gradient Richardson number and the layer Richardson number vary dramatically from around zero to several hundred from spring to neap tides. Stratification and turbulence damping effects near the ETM area induce the upper half (near water surface) of the water body to be ebb-dominant and the lower part (near-bed) to be flood-dominant. These data reveal that the residual pattern of currents, salt flux and sediment flux are non-similar in a stratified estuary, and that **the salinity-induced baroclinic pressure gradient is a major factor controlling the variation of vertical velocity structure.** In addition, field observations indicate that the residual transport generated by internal tidal asymmetry plays a dominant role in maintaining a stable density stratification interface near the estuarine front.

In *Chapter 5*, by means of an improved apparatus, settling velocity (SV) of the Yangtze estuarine fine sediments was studied in the laboratory. The experimental data show that (i) SSC, salinity and temperature all affect SV, but to different extents; (ii) the relationships between SV of estuarine fine sediments and its controlling factors are highly dependent on specific environmental conditions; (iii) **the dependencies of various determinants (SSC salinity and temperature) on SV in different flocculation stages are varied**; and (iv) for the Yangtze estuarine mud, the SV peaks when the SSC is in the range 3-8 kg/m^3, and the salinities for maximum flocculation settling are approximately 7 and 10 psu in dry and wet seasons, respectively.

Chapter 6 attempts to explore the feedbacks of the micro-scale physical processes on the fine sediment dynamics within the river plume of the Yangtze Estuary. Through a numerical sensitivity analysis based on a three-dimensional (3D) small domain model, the effects of micro physical sediment processes related to flocculation and stratification are tested. (i) Settling velocity is a sensitive parameter determining the overall entrained and suspended sediment in the water column. The effect of flocculation on settling velocity controls longitudinal ETM dynamics. (ii) Saltwater intrusion in an estuary firstly creates longitudinal density gradient, which introducing a baroclinic effect. The direction of baroclinic pressure gradient forcing is landward mostly, therefore when the current velocity is relative small during slack waters and neap tides, the advective function is comparable to the baroclinic pressure gradient forcing in the water column (especially near the river-bed), so the internal flow structure will be altered largely. (iii) Furthermore, once the internal flow structure is changed, the up-estuary flux of sediment and salinity will be enhanced near the bottom. Thus the accumulation of denser materials near the convergent area (salt front) will result in a density stratification phenomenon. The vertical density gradient of in stratified flow produces a buoyancy effect on turbulence. This effect suppresses the vertical mixing of momentum and materials (turbulence damping). Then it favors forming of high-concentration layer near the bottom. Meanwhile, settling velocity decreases sharply with an increasing SSC near the bottom, therefore the hindered settling effect enhances the bottom turbidity and promotes the density stratification (including sediment and salinity density). Then again, **sharper stratification will induce stronger turbulence damping and higher bottom SSC. The abovementioned snow ball effect is the primary micro mechanism of fine sediment dynamics in an estuary**. (iv) In addition, with an increasing SSC near the bottom, local bed resistance will be decreased due to drag reduction, tidal stirring is amplified and the erosion capacity is enhanced accordingly. Therefore **the aforementioned snow ball effect might be terminated under the competition between stratification and high-turbidity induced tidal amplification**.

Chapter 7 applied the above findings into the large domain model of the whole Yangtze Estuary and investigated the typical effect of seasonally varying river discharge, wind climate and mean sea level on the seasonal variation of ETM. From observation and modeling data, we concluded evidently that (i) **Both tidal energy and density stratification enhance saltwater intrusion**; (ii) Four independent factors (river flow, wind, mean sea level and water temperature) determining the seasonal sediment regime are identified; (iii) River discharge impacts the pattern of residual currents; (iv) Seasonally varying wind effect alters the longshore currents; and (v) Seasonally varying mean sea level affects the saltwater intrusion length in the DNC.

In this study, in short we highlight that **multiscale physical processes jointly characterize the current and sediment regime in a fine sediment estuarine system**.

KEYWORDS: fine sediment; settling velocity; stratification; flocculation; turbulence damping; modeling; observation; laboratorial experiment; navigational channel; Yangtze Estuary; China

Samenvatting

Estuaria zijn van nature hoogdynamische, snel veranderende systemen, die een complexe combinatie van fysische processen op een aantal verschillende tijdschalen bevatten. De fysische processen rond fijn sediment genieten een toenemende belangstelling van kustwaterbouwkundigen. Eén reden is dat steeds meer landaanwinning nodig is voor een duurzame groei van samenlevingen en tevens meer havens en scheepvaartgeulen. Deze worden in toenemende mate aangelegd in zulke fijn-sediment omgevingen, die voorheen gezien werden als ongunstig voor vaarwegontwikkeling, vanwege de grote kans op sedimentatieproblemen. Een andere reden is dat we geïnteresseerd zijn in het fascinerende karakter van fijn-sediment dynamica.

Het Yangte estuarium is een uistekend voorbeeld van een fijn-sediment estuarien systeem met een matige getijslag (~1-5 m) en een sterke seizoensfluctuatie (~10.000-50.000 m^3/s) in het rivierdebiet. Sedimentatie wordt op dit moment gezien als een belangrijke kwestie. Voordat maatregelen om geulaanslibbing te verminderen kunnen worden ontworpen moeten de redenen en mechanismen ervan worden onderzocht. Daarom wordt het begrijpen van de onderliggende mechanismen rond fijn-sediment transport, de ETM (estuariene turbiditeits maximum) dynamica en sedimentvang in het Yangtze estuarium gezien als een majeure uitdaging bij het onderhouden van de "Gouden Waterweg".

De voornaamste inhoud en conclusies van deze systematische studie naar multi-schaal processen van fijn sediment in een gematigd-getij, convergerend estuariu kunnen als volgt worden samengevat.

In Hoofdstuk 1 wordt de **complexiteit van fijn sediment transport** geïdentificeerd. Afgezien van de interacties tussen rivierafvoer, oceaangetij, windgolven, de Coriolis kracht, zoutwater-indringing en bodemweerstand in een estuarien systeem, tonen micro-schaal effecten van flocculatie, gehinderde bezinken, barocliene forcering, turbulentiedemping en vermindering van de weerstand duidelijke invloed op de interne structuren van stroming, saliniteit en suspensief sediment concentratie (SSC), die op hun beurt weer een effect hebben op de macro-schaal stromings- en sediment regimes en de morfologische ontwikkeling. Na het belichten van de unieke karaktertrekken en uitdagingen van het Yangtze Estuarium worden de de voornaamste doelstelling en de organisatie van deze studie geïntroduceerd.

Hoofdstuk 2 focust op de recente hydrodynamische evolutie van het hele Yangtze Estuarium op de tijdschaal van decaden, afgeleid uit hydrologische waarnemingsreeksen. Gedurende de laatste jaren heeft het systeem menselijke ingrepen ten behoeve van de ontginning van natuurlijke rijkdommen ondergaan op een tot dan toe ongeziene schaal. We hebben vastgesteld dat (i) het waterniveau in de hoofdtak van het Yangtze Estuarium is toegenomen tussen 1998 en 2009; deze toename werd veroorzaakt door veranderingen in het afvoerregime van de volledige rivier (met inbegrip van natuurlijke morfodynamische processen en de topografische terugkoppeling door extreme weerfenomenen en menselijke activiteit); (ii) de afname van het afvoeraandeel ter hoogte van de derde vertakking is een direct gevolg van het Deepwater Navigatie Channel (DNC) project en de daarbij horende morfologische veranderingen in the Noordelijke Doorgang; en (iii) de gradiënten in estuariene omgevingsvariabelen (zoutgehalte en concentraties van suspensiemateriaal) werden ruimtelijk gecomprimeerd, en de zout-zoetovergang werd scherper. **Dit alles heeft het stroomopwaarts opslibben van de vaargeul als indirect gevolg: het stratificatieeffect nabij het turbiditeitsmaximum wordt versterkt, evenals de tendens van stroomopwaarts sedimenttransport.**

In Hoofdstuk 3 worden observaties van de stormgerelateerde vloeistof modder dynamiek in het DNC van het Yangtze Estuarium beschreven, zoals waargenomen tussen Oktober en December 2010 tijdens de passage van een koudefront. De verzamelde gegevens beschrijven hoe net na de storm een grote hoeveelheid fijn gesuspendeerd materiaal in de vaargeul werd gevangen onder de vorm van vloeistof modder. The mechanismen en transportprocessen gelinkt aan de stormgeïnduceerde vloeistof modder-vorming werden geanalyseerd en gepresenteerd vanuit het standpunt van zowel de procesgebaseerde als de ingenieursbenadering. **Met behulp van getij- en golfmodellering kan worden aangetoond dat het stormgerelateerde gedrag van vloeistof modder hoofdzakelijk afhankelijk is van de grootschalige hydrodynamica en de uitwisseling van fijn sediment tussen vaargeul en intergetijdeplaten als gevolg van opwoeling door stormgolven. Dit sediment verzamelt zich in de vorm van vloeistof modder, oscilleert mee met het getij, en hoopt zich op die plaatsen op waar de longitudinale reststromen in de rivier en het tijgedomineerde estuarium relatief zwakker zijn.** Bovendien heeft het hellingafwaarts transport van vloeistof modder het in deze studie geobserveerde event versterkt en verergerd. Onze modelresultaten en observaties tonen het volgende aan: (i) het transport van vloeistof modder is een advectief verschijnsel dat de centrale positie van de fluid mudlaag in de vaargeul bepaalt, en bovendien is het een tijgedreven fenomeen dat de erosie en accumulatie van fluid mud controleert; (ii) zowel de beschikbaarheid van gesuspendeerd materiaal als het lokale reststromingsregime spelen een kritieke rol bij het bepalen van de waarschijnlijkheid dat sediment wordt ingevangen en leidt tot de vorming van vloeistof modder.

In Hoofdstuk 4 worden de in ruimte en tijd variërende meetgegevens geanalyseerd die de spring-doodtijveranderlijkheid van stroomsnelheden, zoutgehalte en SSC in het DNC beschrijven zoals gemeten in het natte seizoen van 2012. Deze gegevens werden voor het eerst verzameld in het midden van het DNC, en naar blijkt documenteren ze een relatief stabiele dichtheidsstratificatie en zouttong, in het bijzonder gedurende doodtij en tijkentering. De convergentiezone van reststromen, zout en sediment tijdens dood- en springtij verplaatst zich tussen het middelste en het zeewaartse deel van het DNC. Het bevordert de vorming van een laag met bijzonder hoge sedimentconcentraties nabij de bodem, wat aanslibbing van de gebaggerde vaargeul veroorzaakt. Zowel het Richardsongetal voor de bodemlaaggradiënt als het Richardsongetal voor gelaagdheid variëren dramatisch, van nul tot ordegrootte honderden naargelang spring- of doodtij. Stratificatie- en turbulentiedempingseffecten in de zone van het ETM leiden tot ebdominantie in de bovenste helft van de waterkolom, en vloeddominantie in de onderste helft. De meetgegevens tonen aan dat het patroon van reststromingen, zout- en sedimentflux niet gelijklopend zijn in een gestratificeerd estuarium, en dat de **zoutgeïnduceerde barocliene drukgradiënt een belangrijke factor is in de variatie van de verticale snelheidsstructuur** Bovendien tonen meetgegevens aan dat de residuele transporten als gevolg van interne getijdenasymmetrie een dominante rol spelen in het behouden van een stabiele dichtheidsstratificatie in de zone van het estuariene front.

In Hoofdstuk 5 werd de valsnelheid van fijne sedimenten uit het Yangtze Estuarium in het laboratorium bestudeerd met behulp van een verbeterd meettoestel. De experimentele gegevens tonen aan dat (i) SSC, zout en temperatuur in verschillende mate de valsnelheid beïnvloeden; (ii) dat het verband tussen de valsnelheid van fijne estuariene sedimenten en de controlerende factoren in hoge mate afhankelijk is van specifieke lokale condities; (iii) de afhankelijkheid van de valsnelheid van de bepalende factoren (SSC, zout, temperatuur) varieert volgens de mate van flocculatie; en (iv) voor wat estuarien slib uit de Yangtze betreft, de valsnelheid maximaal is wanneer de SSC van grootteorde 3-8 kg/m^3 is. De zoutgehaltes voor maximale neerslag van gefloccculeerd sediment zijn ongeveer 7 en 10 psu, respectievelijk tijdens het droge en natte seizoen.

Hoofdstuk 6 tracht de terugkoppeling te karakteriseren tussen de fysische processen die spelen op microschaal en de dynamiek van het fijne sediment in de rivierpluim van het Yangtze estuarium. Met behulp van een numerieke gevoeligheidsanalyse op basis van een driedimensionaal model op relatief kleine ruimtelijke schaal, werden de effecten van fysische sedimentprocessen op microschaal, gerelateerd aan flocculatie en stratificatie getest. (i) Valsnelheid is een gevoelige parameter die de globale hoeveelheid meegevoerd en gesuspendeerd materiaal bepaalt. Het flocculatie-effect op de valsnelheid controleert de longitudinale ETM dynamiek. (ii) Zoutwaterindringing in een estuarium genereert vooreerst een longitudinale dichtheidsgradiënt, hetgeen een baroclien effect introduceert. De richting van de krachtenwerking gelinkt aan de barocliene drukgradiënt is meestal landwaarts. Om deze reden is bij relatief lage stroomsnelheden tijdens kentering en doodtij de advectie van dezelfde grootteorde als de forcering door de barocliene drukgradiënt in de waterkolom. Dit heeft een grote verandering in de interne structuur van de stroming tot gevolg. (iii) Bovendien geldt dat, eens de interne structuur van de stroming is veranderd, de landwaartse flux van zout en sediment nabij de bodem wordt versterkt. Dus de accumulatie van dichter materiaal in de convergentiezone geassocieerd aan het zoutfront zal leiden tot dichtheidsstratificatie. De verticale dichtheidsgradiënt in gestratificeerde stroming resulteert in een drijfvermogen effect op de turbulentie. Dit effect onderdrukt de verticale menging van impuls en materie (turbulentiedemping). In dat geval bevordert het de vorming van een laag met hoge concentraties nabij de bodem. Terzelfdertijd neemt de valsnelheid scherp af met een verhoogde SSC, waardoor hindered settling effecten de bodemlaagturbiditeit versterken en de dichtheidsstratificatie doen toenemen (concentratie van sediment en zout inbegrepen). Een duidelijkere stratificatie leidt dus tot een grotere demping van turbulentie en een hogere SSC nabij de bodem. Dit sneeuwbaleffect is het voornaamste micro-mechanisme voor de dynamiek van fijn sediment in een estuarium. (iv) Bovendien zal door weerstandsreductie de lokale bodemwrijving afnemen wanneer de SSC nabij de bodem toeneemt, getijgerelateerde menging wordt versterkt, en het potentieel voor erosie neemt toe. Het vernoemde **sneeuwbaleffect kan daarom effectief gestopt worden door de tegenwerking tussen stratificatie en de versterkte tijinvloed als gevolg van hoge turbiditeit.**

Hoofdstuk 7 past deze bevindingen toe in een model dat het hele Yangtze estuarium omvat. Het onderzoekt het typische effect van seizoensgebonden rivierafvoer, windklimaat en gemiddeld zeeniveau op de veranderlijkheid van het ETM. Uit meetgegevens en modelresultaten kan besloten worden dat (i) **zowel getijslag als dichtheidsstratificatie bevorderen zoutwaterindringing;** (ii) Vier onafhankelijke factoren (rivierafvoer, wind, gemiddeld zeeniveau en watertemperatuur) bepaling het seizoensgebonden sedimentatiepatroon; (iii) Rivierafvoer beïnvloedt het residueel stromingspatroon; (iv) Seizoensgebonden variatie in windrichting en -snelheid veranderen de kustlangse stroming; en (v) Seizoensgebonden veranderingen in de gemiddelde waterstand beïnvloeden de afstand waarover zoutindringing plaatsvindt in het DNC.

Samengevat tonen we aan in deze studie hoe **fysische processen op meerdere schalen samen het stromings- en sedimentatiepatroon bepalen in een estuarien systeem met fijn sediment.**

TREFWOORDEN: fijn sediment; valsnelheid; stratificatie; flocculatie; turbulentiedemping; modelleren; observaties; laboratoriumexperimenten; vaargeul; Yangtze Estuarium; China

* This summary is translated from English to Dutch by Mr. Johan Reyns, but the author himselft is responsible for the accuracy.

Acknowledgements

It was indeed a value-added trip for me to do the PhD research in the past 5 years, though I led a busy life and I had to travel between Delft and Shanghai for my PhD study and consultant research jobs, respectively. There were painfully tough moments, which turned to be extremely rewarding when this thesis was completed. This trip has provided me with academic experience that was challenging, as well as an opportunity to see various thoughts, ways of life, cultures, and dreams. I would like to express my sincere gratitude to all the people that I met during the trip, and especially to those who have made direct or indirect contribution to this study.

Thank you to my family, including my wife Xu Yan, my parents in law, my mother and my father, who always show great tolerance and love to deal with family matters and take care of my son Wan Jinhui when I was absent. They present to me a "complaint free" attitude, which encourages me to keep smile to everything.

Prof. Dano Roelvink is gratefully acknowledged for his insightful advice, helpful guidance and artistic inspiration to this study. I thank you from the bottom of my heart for allowing me to come and the nice experience. His modeling philosophy *(Roelvink and Reniers, 2012)* has reshaped my understanding of sediment transport and morphodynamics. It is my honor to be his PhD student. Thanks ever so much for the fun, your sense of humor, and your patience. I have felt so motivated after each meeting with you. Moreover, you encouraged me to think independently, you taught me to do writing with simple words and trained me to express academic ideas in gentle and moderate way.

Mr. Gao Min is greatly thanked for having recommended me to do this PhD research. In 2008, this guy did a very impressive study on the topic of sediment transport and morphological process in the Yangtze Estuary *(Gao, 2008)*. It is just this research that attracted Dano's great interest on the Yangtze estuarine sediment dynamics. With the funding of the project, Research on Sediment from Upstream to Estuary (ReSedUE), Prof. Dano brought me to Delft to pursuit the beauty of the Yangtze Estuary.

Special thanks go to my colleagues (Wu Hualin, Qi Dingman, Gu Fengfeng, Kong Lingshuang, Wang Wei, Shen Qi, Liu Jie, Liu Gaofeng, Wang Yuanye, Le Jiahai, etc.) at Shanghai Estuarine Coastal Science Research Center and all the members of the Yangtze Estuary Research Team, under the management of the Yangtze Estuary Waterway Administration Bureau. Without their contribution and support, I cannot have the chance to access these experimental facilities, measurement data and numerical models.

It was fantastic to meet all of you, colleagues and PhDers in IHE, Rosh, Mick, Ali, Johan, Jolanda, Tonneke, Wendy, Anique, Sylvia, Peter, etc. Thank you for your help during my study in Holland.

The scientific discussions and personal communications with Han Winterwerp, Wang Z.B., Jin Liu, Robert, Shi Wei, Swart, J.Z. Shi, Zhu Jianrong, Cheng Peng, Yu Qian, Ma Gangfeng, Han Yufang, Chu Ao, Cheng Wenlong, Ye Qinghua, Shao Yuyang and Wang Li inspired me and brought me some new ideas.

Prof. Richard Burrows is thanked for having sent some copies of literatures about fluid mud. Lu Shengzhong, Chen Xi and Huang Wei are thanked for their enthusiastic assistances on the laboratory experiments in this study. My BSc supervisor Prof. Yu Minghui and MSc supervisor Prof. Dong Yaohua are acknowledged for their continuous encouragements.

The hot pot, BBQ and marathon friends in Holland and China are appreciated for sharing the happy and relaxing moments together. They are Yang Zhi, Li Shengyang, Wang Chunqin, Guo Leicheng, Yan Kun, Pan Quan, Lin Yuqing, Xu Zhen, Chen Qiuhan, Chu Kai, Zhao Gensheng, Zhang Yong, Ouyang Xiaowei, Fu Bingjie, Zhao Dezhao, Wang Wei, Shen Qi, Pan Jiajun, Dong Bingjiang, Zhou Chi, Zuo Liqin, Wang Hao, Li Shouqian, etc.

I would like to thank Prof. John Z. Shi for his constructive comments and improvements.

I thank the developers and contributors of the Delft3D, GOTM, SELFE/ELCIRC, FVCOM, NaoTide for the open access to source codes.

WAN Yuanyang
Shanghai, China
April, 2015

Contents

Introduction

Highlights

(1) The complexity of fine sediment transport in estuaries is identified.

(2) The complexity of fine sediment dynamics in the Yangtze Estuary is distinguished.

(3) The main content and organization of this study are presented.

1.1 Background

As we all know, plants, animals and mankind need water. Ever since the dawn of civilization, rivers have been regarded as the center of human activities. Most existing estuaries were formed during the Holocene epoch by the flooding of river-eroded or glacially scoured valleys when the sea level began to rise about 10,000-12,000 years ago *(Wolanski, 2007)*. More than 60% of the world's population lives along estuaries. Most important cities in the world are located along the banks of estuaries, such as Shanghai (Yangtze Estuary), New York (Hudson Estuary), Rotterdam and Antwerp (Scheldt Estuary), Hamburg (Elbe Estuary), Liverpool (Mersey Estuary), New Orleans (Mississippi Estuary), Cairo (Nile Estuary) and Vancouver (Fraser Estuary). Understanding physical processes of an estuarine system is of crucial importance to human social development. The dynamics of sediment transport is amongst the most important variable characteristics of an estuary.

In estuaries, sediment transport and morphodynamics always create great challenges for human beings to utilize a wide variety of natural gifts and also pose challenges to delta social and economic life. The interconnected sediment physical processes of erosion, transportation and deposition in estuaries are more unpredictable and uncontrollable than those in rivers. The difficulty and complexity are mainly due to the changing conditions and surroundings (see *Table 1-1*), when those sediments travel from river to estuarine zone, especially for the fine-grained sediments (we call it "fine sediment" in short) in an estuarine turbidity maximum (ETM) zone.

Table 1-1. The difference between river and estuary, related to hydrodynamics and sediment transport[1].

Location	River	Estuary
Control force	Gravity force Discharge Bed resistance Precipitation an evaporation Human modification	Astronomic tides Gravity force Freshwater inflow Bed resistance Coriolis force Wind effect Wave function Longshore current Human modification
Hazard and risk	Flood	Storm (typhoon or tsunami) Flood Seal level rise (global warming)
Current direction	unidirectional flow (normally from upstream to downstream)	Reciprocating current (to-and-fro flow) or rotational flow
Special physical process and phenomenon	Debris flow Density current Bed load transport[2]	Estuarine turbidity maximum (ETM) Fluid mud Turbulence damping Drag reduction Flocculation Saltwater intrusion Stratification
Geometry	Meandering Straight Dam-controlled Dike-protected Bifurcated	Alluvial fan, bird's foot or dendritic shape *(Seybold et al., 2007)*

Note: 1. Some of these differences are site- and conditional- specified, i.e., in some estuaries, flood is not an effective risk *(Su and Wang, 1989)*; 2. Comparatively, suspended sediment transport is the dominant mode of transport in estuaries generally *(van Rijn, 2007)*.

Since the sediment composition and cohesive content will change the degree of cohesion, the rigid classification of fine and coarse sediment is not defined in traditional textbook.

Generally, when the median grain size (D$_{50}$) of natural sediment is finer than 62 μm *(Mehta and McAnally, 2008)*, the flow-sediment interactions (micro-scale physical processes) will impact macro-scale behavior of hydrodynamics and sediment transport. That is much similar to "micro-scale quantitative accumulation leads to macro-scale qualitative transformation". Thus, the constituent processes and control forcing of hydrodynamics, sediment transport and morphodynamics in a fine sediment estuarine system are not just related to those macro-scale forcing, such as riverine inflow, marine tide, bed resistance, the Coriolis force, wind effect, wave function and longshore current *(Chang and Isobe, 2003)*.

With regard to flow-sediment interactions, many scientists and engineers have contributed to topics that are closely associated with fine sediment dynamics *(Winterwerp, 1999; Le Hir et al., 2011; Mehta, 2014)*. During the past engineering applications and practices in the world, including port and harbor projects, land reclamation, dam and reservoir constructions, navigational channel development, environmental protection and disaster prevention, many effective experiences and fundamental understanding have been achieved.

(1) With their cohesive nature, fine sediments are prone to aggregation and formation of flocculated network structures (flocs) during the settling process *(Kineke and Sternberg, 1989; van Leussen, 1994)*, introducing distinct flocculation acceleration and hindered settling phenomenon (flocculation settling velocity effects);

(2) The presence of salt and suspended particulate matter (SPM) in water column increases mixture density *(Qiu et al., 1988; Jay and Smith, 1990; Jay and Musiak, 1994)*, resulting in longitudinal density variations and baroclinic pressure gradient forcing (baroclinicity effect);

(3) The vertical gradient of fluid density creates a stratified water column and introduces buoyancy effect on turbulence *(Winterwerp, 2011b)*, damping the flow turbulence (turbulent damping effect);

(4) The SSC- and salinity- induced stratification produces a stratified bottom boundary layer *(Li and Gust, 2000)*, altering the logarithmic velocity profile near the bottom and reducing drag coefficient (drag reduction effect).

The above effects govern the vertical distributions of turbulence, SSC, salinity and current directly. Moreover, they will jointly be coupled with those macro-scale forces, inducing tidal asymmetry, lag effects, tidal pumping and landward residual circulation. They are the primary physical mechanisms responsible for sediment trapping, harbor siltation, waterway back-silting, and formations of salt wedge, fluid mud and ETM in many estuaries *(Burchard and Baumert, 1998; de Nijs and Pietrzak, 2012; Wan et al., 2014a)*.

Many studies (e.g. *de Nijs and Pietrzak, 2012; van Maren and Winterwerp, 2013*) highlighted the importance of these effects on model performance in simulating the fine sediment transport and morphodynamic evolution, and also on the understanding of many related phenomenon. These effects are considered as the substantial difference between sand and mud models.

For modeling techniques, the flocculation settling velocity effect is commonly involved via formulating an empirical or experimental SSC-dependent settling velocity *(Guan, 2003; Song and Wang, 2013)*. The baroclinic effect is mainly taken into account by computing the baroclinic pressure gradient terms in the horizontal momentum equation *(DHI, 2009; de Nijs, 2012)*. The uncertainty of modeling of fine sediment dynamics is partly due to the latter two effects. Both turbulent damping and drag reduction are related to density stratification in the water column; their relationship is still unclear due to our poor understanding of turbulence and the difficulty in micro-scale observation adjacent to riverbed. Practically in numerical model, turbulence damping effect is either considered through accounting for the effect of buoyancy on the production and dissipation of turbulent kinetic energy *(Deltares, 2014)* using a second-order (or two-equation) turbulence model (i.e. the standard k-ε model); or it is represented by an empirical parameterization of the turbulent Prandtl number and gradient Richardson number in a zero-order turbulence model *(Vasil'ev et al., 2011)*. Drag reduction phenomenon has been observed in field and laboratory experiments *(Gust and Walger, 1976; Wang et al., 1998; Cheng et al., 1999; Wang, 2002)*, and an improved formula about the bottom friction coefficient accounting for drag reduction effect was introduced in modeling *(Wang, 2002)*.

The roles of each micro-scale physical process and macro-scale control force governing fine sediment dynamics in an estuarine system are very essential for us to understand the behaviors of natural estuarine process and their feedback to human intervention. Many quantitative and qualitative studies have been carried out to confirm one or two effects on hydrodynamics and suspended sediment transport by field observation, idealized model, analytical analysis and process-based model. *Krone (1962), Ali and Geoprgiadis (1991)* and *McAnally (2000)* identified special physical characteristics of fine sediment. Settling velocity of fine sediment attracted a wide range of attention in the world *(Owen, 1971; van Leussen, 1994; Winterwerp, 1998; Agrawal and Pottsmith, 2000; Manning and Schoellhamer, 2013)*; SSC, salinity, temperature and turbulent intensity can significantly control settling velocity of fine sediment. *Wolanski et al. (1988)* and *Mehta (1991)* studied the character and behavior of fluid mud. *Winterwerp (1999)* and *Le Hir et al. (2000)* highlighted the existence of the high-concentration suspended sediment in the fine sediment regime. The sediment trapping and ETM formation mechanisms were addressed from various viewpoints: (1) flocculation settling *(Manning et al., 2010)*; (2) internal tidal asymmetry *(Jay and Musiak, 1996)*; (3) external tidal asymmetry/tidal pumping *(Uncles et al., 2006b; Cheng et al., 2011)*; (4) tidal straining *(Simpson et al., 1990)*; (5) turbulence damping *(Geyer, 1993; Winterwerp, 2011b)*; (6) tidal resuspension *(Shi, 2010)*; (7) settling lag *(Postma, 1961)*; (8) drag reduction *(Li and Gust, 2000; Winterwerp et al., 2009)*. In addition, the nonlinear relationship between variations of freshwater inflow and sediment regime was modeled and illustrated *(Uncles et al., 2006; Song and Wang, 2013)*. *Ralston et al. (2010)* and *de Nijs and Pietrzak (2012)* found that the salinity intrusion length is a main parameter controlling fine sediment trapping. *Dronkers (1986)* and *Yu et al. (2014)* thought tide-induced residual transport of fine sediment dominates the sediment trap in estuaries. Wind and wave effects were significantly responsible for sediment resuspension *(Partheniades, 1965; Hu et al., 2009b)*. *Uncles et al. (2002)* analyzed the dependence of estuarine turbidity on tidal intrusion length, tidal range and residence time. *Wang et al. (2014)* found that the flood-induced high concentration event influences tidal amplification and SPM transport. Nevertheless, there is still a need to investigate the role of each effect above in controlling the nature of fine sediment dynamics in a large scale estuarine system. So, a systematical study is needed to access the nature of fine sediment dynamics.

1.2   The Yangtze Estuary

In this study, we choose the Yangtze Estuary as the study case. The Yangtze Estuary is one of the most charming estuaries as related it complexity and challenge on fine sediment transport. After the implementation of the Three Gorges Dam project (TGD, *Figure 1-1a*) since 1994 and the -12.5m Deepwater Navigational Channel project (DNC, *Figure 1-1b*) since 1998 in the Changjiang River basin, some critical concerns *(e.g. Wu et al., 2003; Shen and Xie, 2004; Xu and Milliman, 2009; Shi et al., 2012; Dai et al., 2013)* over the environmental and sedimentation issues have been aroused by many people. It is controversial that the sediment discharge from the Yangtze into the sea decreased apparently *(Yang et al., 2014)*, however at the same time the sedimentation in the DNC increased obviously *(Liu et al., 2011)*, see *Figures 1-2* and *1-3*.

Since the 1950s, a wide range of studies *(e.g. Chen, 1957; Shi et al., 1985; PDC, 1986; Wang, 1989; Shen and Zhang, 1992; Chen et al., 2001a; Gao, 2007; Zhu et al., 2014)* have been conducted to explore the physical processes of fine sediment and attempt to develop a sea-going navigable channel in the estuarine area. However, it is still challenging about the fundamental processes and mechanisms responsible for flow-sediment interaction, sediment trapping, dynamics of ETM and fluid mud *(Shi, 2010; Song et al., 2013)*. There are a number of uncertainties and problems for engineering application in this area. At the beginning, the complexity and recent challenge in this study case are identified and demonstrated.

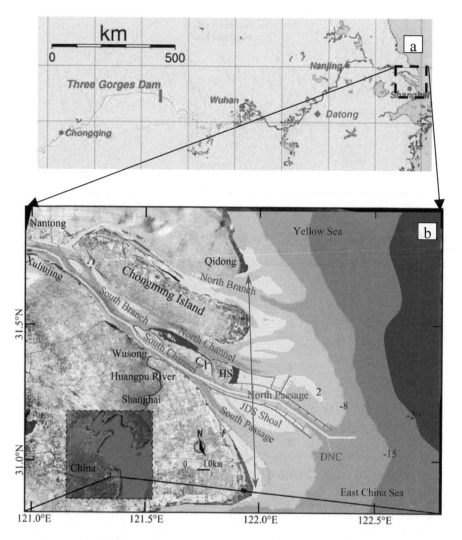

Figure 1-1. (a) Sketch of the Yangtze River, Datong is tidal range limit of the Yangtze Estuary (after Wikipedia). (b) General layout of the Yangtze Estuary. CX, HS, JDS, QR and DNS are the abbreviations of the Changxing Island, Hengsha Island, Jiuduansha, Qingcaosha Reservoir and deepwater navigational channel, respectively. The yellow bold line is the -12.5 m deepwater navigation channel; the purple lines denote the reclaimed lands; the black lines in the North Passage are the groins and dikes of the DNC project. The purple arrowed line shows the width of the mouth of the estuary.

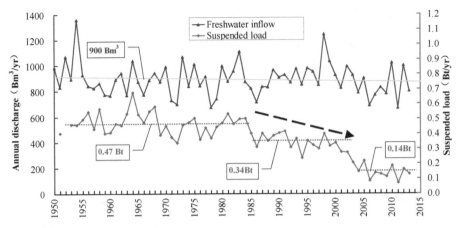

Figure 1-2. Time series of annual freshwater inflow and suspended load at the Datong station from 1950 to 2013 (date source: *CWRC, 2013*)[1]. The overall decreasing trend of suspended load is indicated by the black dashed arrow, while the multi-year-averaged suspended load is by the purple dotted line.

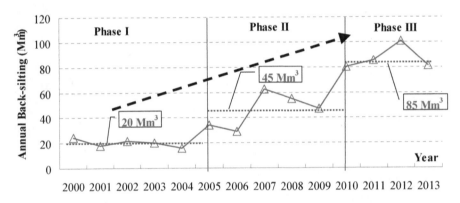

Figure 1-3. Time series of annual quantity of back-silting in the DNC from 2000 to 2013. The overall increasing trend of back-silting is indicated by the black dashed arrow, while the phase-averaged back-silting is by the purple dotted line. The engineering process related Phase I-III refers to *Figure 2-2*. The dredging maintenance range along the DNC in the period of 2000-2004 (Phase I), 2005-2009 (Phase II) and 2010-present (Phase III) was Cells A-Z, Cells IIN-A~IIW-B and IIIA~III-I (cell locations see *Figure 1-10*), respectively. It is worth noting that the maintenance water depth increased from 8m (Phase I), 10m (Phase II) to 12.5 m (Phase III).

1.2.1 Complexity

The hydrodynamic, sedimentological and geomorphologic characteristics about the Yangtze Estuary have been introduced and investigated by many people *(Chen et al., 1985; Hu et al., 2009b; Shi, 2010; Wu et al., 2012; Jiang et al., 2013b; Maren et al., 2013; Song et al., 2013; Wang et al., 2013)*. But the complexity and difficulty of the Yangtze Estuary related problem are always easily underestimated or overlooked. As one of the biggest estuary in the world, in compare with the other estuaries (see *Table 1-2*), the complexity mainly associated to the hydrodynamics and fine sediment transport in the Yangtze Estuary is due to the huge and

---

[1] Online available: *www.cjw.com.cn/zwzc/bmgb/nsgb/201503/t20150316_135210.html* (in Chinese)

6

complex estuarine geometry and conditional variation of ETM dynamics.

Table 1-2.　Comparisons of fine sediment estuaries in the world.

Estuaries	Basin area (10^3km^2)	Width at the mouth (km)	Tidal range (m)	Tidal limit (km)	Yearly freshwater discharge (m^3/s)	Geometry (L.Brennan et al., 2002)	Classification related to forcing
Yangtze[1]	1,808	90	0.2-4.6	650	30,166	Dendritic shape Multi-channel	River- and tide-dominated
Scheldt[2]	21.8	5	2-6	156	120	funnel shape Multi-channel	Tide-dominated
Mississippi[3]	3,268	2	0-1	250	18,400	bird's foot shape Multi-channel	River-dominated
Amazon[4]	7,050	150	1-11	1,000	209,000	Dendritic shape Multi-channel	River- and tide-dominated
Hudson[5]	36.3	3	0.8-1.8	250	620	funnel shape with a inner bay Single-channel	River-dominated
Severn[6]	11.4	35	2-15	250	61	Humpback shape Single-channel	Tide-dominated
Elbe[7]	148	15	2-5	120	711	funnel shape Single-channel	Tide-dominated
Gironde[8]	110	20	2-5	150	1,100	Humpback shape Single-channel	Tide-dominated

Note: 1. the data are refereed from Google Earth, Wikipedia (*en.wikipedia.org*) and *(Chen et al., 1999; Shi, 2010; Zhang et al., 2012b)*; 2. from *(Middelburg et al., 2002; Meire et al., 2005; van Kessel et al., 2011a)*; 3. from *(Penland and Suter, 1989; Corbett et al., 2004)* ; 4. from *(Kineke and Sternberg, 1995; Geyer et al., 1996; Vinzon and Mehta, 2003)*; 5. from *(Nepf and Geyer, 1996; Levinton and Waldman, 2006)* ; 6. from *(Kirby and Parker, 1983; Manning et al., 2010)*; 7. from *(Puls et al., 1988; Abril et al., 2002)*; 8. from *(Uncles et al., 2002; van Maanen and Sottolichio, 2013)*.

1.2.1.1　Huge and complex estuarine geometry

The natural geometry of an estuary is usually more complicated than that of a river. It is jointly governed by forcing conditions from the Coriolis force, flow, wind, and wave, the availability and erodibility of bed material, human activity and modification, and the extreme events such as flood, storm, typhoon and earthquake. The two remarkable features of the Yangtze Estuary are highlighted as follows.

(1) Huge size

The width at the mouth of the Yangtze Estuary is about 90 km (see the purple arrowed line in *Figure 1-1b*), and the landward tide can propagate to 600-700 km far away from the mouth. Comparing the Yangtze Estuary with the Western Scheldt estuary, the Elbe estuary and the Gironde estuary (see *Figure 1-4* and *Table 1-2*) with regard to their geometry, we can see that the size of the Yangtze Estuary is much greater than the other three estuaries. It can be expected that the horizontal distributions of grain size *(Li et al., 2008)*, tidal prism, hydrodynamics, SSC, wind and wave climate and geomorphologic conditions are of wide ranges and high variations. In addition, multi- time and space scales physical processes (from turbulent eddy to sea-level-rise) *(de Vriend, 1991)* are all included in and works in the huge size estuarine system. Therefore each step of simplification of the complex system may lead to one uncertainty and raise a new issue. And more, the big scale of the estuary is more costly (in time, space and computation) and difficult to conduct a relatively systematic field survey, physical and numerical modeling than others.

On the other hand, the Yangtze Estuary is significantly enforced by riverine inflow and tidal force jointly. The maximum and minimum run-off at the Datong station is 92600 m^3/s and 4620 m^3/s, respectively. The seasonal run-off difference is more than 20 times. The maximum and minimum tidal range is 0.2 m and 4.6 m, and the difference is 23 times. The flexible combinations of the two above control forces in the huge estuary result in a wide range variation of hydrodynamic condition.

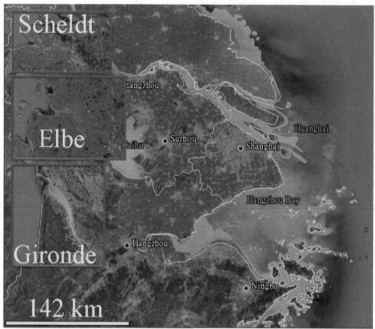

Figure 1-4. Comparison of the size of different estuaries (Yangtze, Western Scheldt, Elbe and Gironde) with the same length scale (based on Google Earth).

(2) Three order of bifurcations

The Yangtze Estuary (*Figure 1-1-b*) is first divided by the Chongming Island into the North Branch and the South Branch (1st bifurcation). The South Branch is subdivided again into the North Channel and the South Channel (2nd bifurcation), and lastly the South Channel is further subdivided into the North Passage and the South Passage (3rd bifurcation). We know that the flow and sediment portion pattern near a river channel junction is commonly unstable *(Kleinhans et al., 2013)* and varied with nearby bed resistance, flow regime and the development of the bifurcation. The currents at a intersection water area, just as the traffic system at a crossing on land, is much complicated and sensitive than that at a simple and straight river or trench, especially at those bidirectional (flood and ebb) channels, where the flood tides or say marine influences, and ebb flows or say fluvial influences interacted frequently. The three bifurcations of Yangtze Estuary are just up to the feature like that, see *Figure 1-1*. In the detailed map of the focused area, it's clear that the ebb flows and flood tides coming from different watercourses join together and exchange with each other.

For the tidal energy enforced system, the ebb and flood portion of current and SSC are not the same, see *Figure 1-5*. On the other hand, the flow and sediment portion at a lower bifurcation will impact the portion at the upper bifurcation. For instance, if the human invention decreases the flow portion at the North Passage (3rd bifurcation), the overall river roughness in the South Channel will be increased, and the portion at the 2nd bifurcation will be adjusted accordingly. Furthering, the flow portion and current regime near the 1st bifurcation also will be varied to a certain extent. Thus, it means the distributary behavior make the entire Yangtze Estuary linked to be a high sensitive and indivisible estuarine system.

In addition, the current impacted by irregular semidiurnal tide and seasonally varying freshwater inflow couples with the Coriolis force and wind waves, and more importantly it interacts with the complex boundary and bathymetry in the Yangtze Estuary. Thus, the velocity will not be primarily determined by the water level gradient (gravitational circulation) in some locations (such as the ETM zone). That is the main difference of current condition between small and big size estuaries.

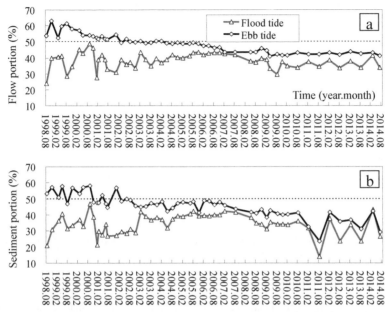

Figure 1-5. Time series of measured flow (a) and sediment (b) portion in the North Passage (in proportion to that in the South Passage) during flood and ebb tides from 1998 to 2014.

(3) Tidal flat and shallow shoal (sediment "bank")

At the four seaward outlets (North Branch, North Channel, North Passage and South Passage) of the estuary, there all exist the mouth bar *(Edmonds and Slingerland, 2007)* or estuarine turbidity maximum. Near these mouth bar zones, a number of movable tidal flats and shoals are located in the shallow areas (see the grey color areas in *Figure 1-1b*). Tidal flats firstly act as "sediment bank" and impact the flood-ebb tidal, spring-neap tidal and seasonal morphological processes in the Yangtze Estuary. Self-adjusting of tidal flats will activate those weak consolidated bed materials and supply enough SPM to adjacent water column. The residence time (water age) *(Deleersnijder et al., 2001)* is relative long due to that tide can reach the Datong (see *Figure 1-1a*). Thus these sediments in the tide-influenced area are of high activity and will oscillate with currents for a long time. Secondly, the bathymetry in the estuary is varied dramatically and the topography is strip-shaped with multiple shoals and trenches. So along- and cross- channel water depths change dramatically, introducing a strong non-linear advection in this estuarine system and enhancing the complexity of the hydrodynamic processes of the Yangtze Estuary.

(4) Daily dredged navigational channel

Due to the siltation characteristics of the DNC (the channel is located in the mouth bar area of the Yangtze Estuary), daily dredging activity is necessary for maintaining it accessibly. For economic reasons, most dredged material currently is disposed of at those disposal sites *(Figure 1-6)*. The average yearly amount of dredged materials in the DNC is ~80 million m^3 from 2010-2014. Only 40% of them are pumped to the nearby reclaimed land (green highlighted area in *Figure 1-6*), and 60% of them need to be deposited at these disposal sites *(Liu et al. 2012)*.

Those disposal sites are located near the DNC, and there is concern that the deposited dredged sediments will return to the channel through the following ways *(Liu et al. 2012)*. Firstly, high-concentration effluent appears once those dredged sediments has been released and deposited in the water, and it produces hyperconcentrated benthic layer or fluid mud and may downslope into the channel directly. Secondly, the sediment plume agitated by the dredging and dumping activities will increase the settling velocity and stratify the suspension, which favors the sediment trapping in the channel indirectly.

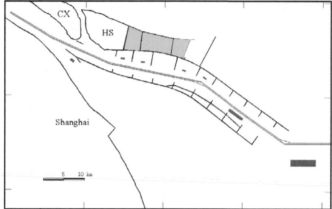

Figure 1-6. Layout of the dredging disposal sites of the DNC. The pink filled areas are the disposal sites and the green is the reclaimed land.

(5) Human intervention

As a place for critical social and economic activities in China, a number of human modifications related to water and land resources utilization have been conducted in the Yangtze Estuary. Three main projects, the QCS Reservoir (location see *Figure 1-1b*), the DNC Project and the Yangshan Port *(Zuo et al., 2012)*, as well as numerous harbors and land reclamations along the Shanghai and Jiangsu coasts, have been constructed in the past decade. These human modifications locally or regionally intervened with the natural hydro- and morpho- dynamic processes of the Yangtze Estuary, and reshaped the corresponding submerged geometry.

1.2.1.2   Conditional variations of ETM dynamics

To understand dynamics of the ETM, 28-h-long mooring station data were obtained (the stations are shown in *Figure 3-9*), which collected hourly, vertical profiles of SSC and salinity, on spring tides both in the wet season (Aug. 12-13) and in the dry season (Jan. 31-Feb. 1) of 2010.

The stations are located on the flank of DNC, and to a reasonable extent cover the longitudinal gradients and vertical stratification of the channel. Tidal period averaged values of salinity and SSC at ebb and flood tides are estimated for wet and dry seasons, respectively *(Figure 1-7)*. Based on these monitoring data, we can see three significant features in the figure:

(1) The vertical salinity distribution in the wet season is strongly stratified, even including a salt wedge during flooding *(Figure 1-7f)*, while in the dry season, it seems vertically well mixed.

(2) Mainly because of an increasing river discharge from dry to wet seasons, the upward intrusive distance of saline water in the dry season is greater than in the wet season, and the 5 psu (practical salinity unit) isoline moves up-estuary at least 30 km from wet season to dry season.

(3) The biggest top-to-bottom difference of SSC is about 3 times in the dry season, while this value in the wet season is more than 20. It indicates the overall SSC in the dry season is much smaller and less stratified than in the wet season.

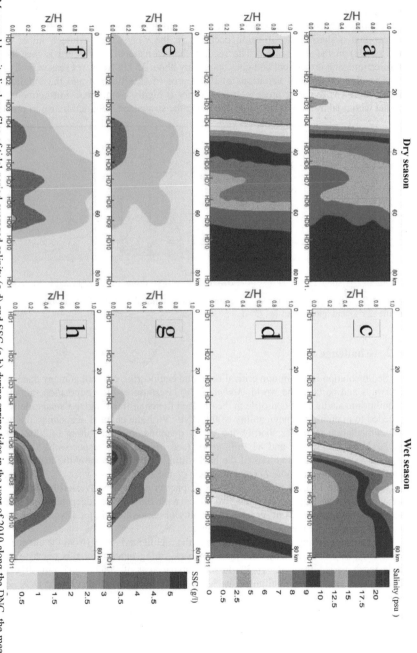

Figure 1-7. Measured longitudinal profiles of tidal period averaged salinity (a-d) and SSC (e-h) during spring tides in the year of 2010 along the DNC, the measurement time in dry and wet seasons are 08-02-2010, and 16-08-2010, respectively. (a c e g) is the salinity and SSC of ebbing time. Locations of the measurement stations HD1-HD11 can be referred *Figure 3-9*. SSC of flooding time during a tidal cycle, (b d f h) is the salinity and

The results imply that the stratification and suspended sediment regime are conditional, determined by different tidal cycles and types, seasons, and locations even in the same estuary; a similar perspective is also described by *Valle-Levinson (2010)*. The stratification parameter (SSP) is employed (detailed description see *Section 4.2.2.1*) to quantify the seasonal variations of stratification effect in the water column. The stratification effect is considered to have a pronounced effect on sediment trapping and vertical sediment mixing.

The tidal- and vertical- averaged SSP in wet and dry seasons are calculated in *Figure 1-8*. It is clear that a sharper stratification occurred in the wet season, while a rather small SSP (well-mixed) in the dry season. The second characteristic of this figure is that the greatest stratification zone is different and goes up-estuary about 35 km from wet season to dry season.

The distinct seasonal variations of the degree of stratification and the longitudinal SSC suggest the ETM dynamic process in the estuary is highly conditional. Furthermore, they will interact with tidal hydrodynamics and determine the suspended sediment transport processes.

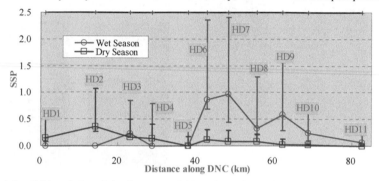

Figure 1-8. Tidal period- and depth- averaged SSPs along the DNC. Vertical lines represent lower quartile (25%) and upper quartile (75%).

1.2.2 Challenge

Sedimentation is a common critical issue and engineering concern at many dams, reservoirs, waterways and ports in the world *(Mehta, 2014)*. Because of high deposition rate and serious unequilibrium status of SSC profile in the dredged trench of the Yangtze Estuary, daily dredging is needed to maintain the navigability of the DNC. With the channel deepening, the quantity of the sedimentation increases sharply, the yearly quantities of back-silting in 2002, 2005, 2008 and 2013 are 21, 36, 57 and 81 million m^3 (see *Figure 1-3*), respectively. In addition, the temporal and spatial distributions of the back-silting in the DNC show the impressive features as follows, according to *Table 1-3* and *Figures 1-9* and *1-10*.

(1) From the Phase I project to Phase III project of the DNC (referring *Figure 1-3*), the annual back-silting volume doubled two times with the channel deepening. At present, the magnitude more than 80 M m^3/yr (annual quantity of back-silting in the DNC) posts tremendous economic pressure for the channel maintenance.

(2) The back-silting is quite centralized at the highlighted reach (from the channel cell H to O, see *Figures 1-9* and *1-10*). The ratios of back-silting quantity at this reach to that of the whole reach varied from 46%, 63% to 85% at the year of 2002, 2005 and 2010.The rapid sedimentation rate at the highlighted reach endangered the navigability of the DNC, especially during July to October.

(3) High siltation rate within the DNC occurs in the wet season (June to November), see *Table 1-3*. The ratio of total siltation quantities in wet and dry season is 86:14.

(4) Comparisons of *Figures 1-7* and *1-9* suggest that the high siltation rate in the wet season and the highlighted reach is associated with seasonally conditional density stratification.

The serious sedimentation problem in the DNC ranks as a key issue in the Yangtze Estuary recently. Before conceiving a measure to mitigate channel siltation, the causes and mechanisms related to the above characteristics of sedimentation should be investigated. Thus, understanding the underlying mechanisms related to sediment trapping, ETM dynamics and fine sediment transport in the Yangtze Estuary are considered as a major challenge to maintain the "golden waterway".

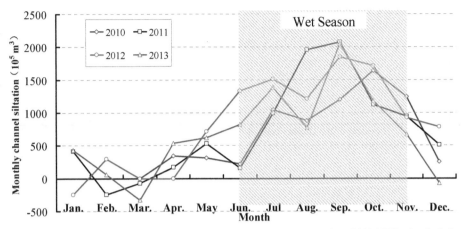

Figure 1-9. Monthly channel siltation in the DNC after the Phase III project (2010-2013), the shaded area means in the wet season.

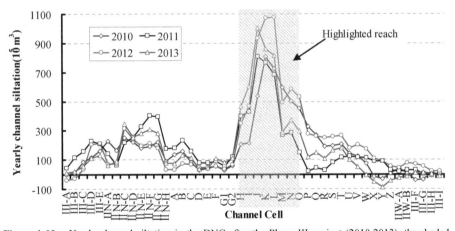

Figure 1-10. Yearly channel siltation in the DNC after the Phase III project (2010-2013), the shaded area means the highlighted reach of high back-silting rate.

Table 1-3. 4-year averaged monthly channel siltation and its proportion of one year in the DNC (2010-2013). Data in black and red refer to the dry season and wet season, respectively.

Month	Jan.	Feb.	Mar.	Apr.	May	Jun.
Siltation ($10^5 m^3$)	98	-62	-52	261	542	633
Proportion (%)	1	-1	-1	3	6	8
Month	Jul.	Aug.	Sep.	Oct.	Nov.	Dec.
Siltation ($10^5 m^3$)	1234	1204	1791	1410	944	368
Proportion (%)	15	14	21	17	11	4

13

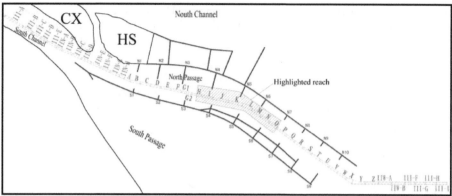

Figure 1-11. Location of the channel cells of the DNC, shaded area denotes the highlighted reach for high back-silting rate.

1.3 Objective

This PhD research has been carried out under the framework of the ReSedUE (**Re**search on **Sed**iment from **U**pstream to **E**stuary, an international partnership project led by Prof. Dano Roelvink), and also supported by the ECSRC. The joint interest is to improve our understanding of the physical processes of fine sediment in the Yangtze Estuary and to motivate new methodology and strategy on numerical modeling of mud transport. The specific proposed objectives (not limited to) are as follows.

(1) To investigate the basic characteristics of the Yangtze Estuary associated to the relationships among hydrodynamics, sediment transport and navigational channel siltation.

(2) To understand the hydrodynamic response to channel deepening process of the North Passage of the Yangtze Estuary.

(3) To study the formation mechanism of a fluid mud event in the DNC.

(4) To elucidate the features of ETM dynamics and saltwater intrusion within the Yangtze river plume.

(5) To measure settling velocity of the Yangtze fine sediment in the laboratory.

(6) To model the vertical structure of current, salinity and SSC.

(7) To evaluate the seasonal variation of fine sediment dynamics in the Yangtze Estuary.

1.4 Outline

This PhD study will be addressed in 8 chapters. Except for the first and final chapters, each chapter is organized in the format of a journal-paper and has the main headings of: INTRODUCTION, METHODS, RESULTS/DISCUSSION, and CONCLUSION. The detailed organization of this thesis is as follows.

Chapter 1 identifies the characteristics of the physical processes of estuarine fine sediment. And the complexity of the study case (the Yangtze Estuary) is highlighted.

Chapter 2 presents a series of measurement data and 2D modeling results to illustrate the hydrodynamic processes in the Yangtze Estuary from 1998 to 2009, showing the feedbacks between morphological and hydrodynamic processes. And the influences of the hydrodynamic evolution on density stratification and sediment transport are discussed.

Chapter 3 provides an investigation on a fluid mud event in 2010. Fluid mud is an extreme transport mode of fine sediments. A fluid mud event may illustrate the fine sediment dynamics and trapping processes clearer than regular condition. So, in this chapter, we investigate the dynamics of fluid mud firstly. The mechanism and transport process of the storm-induced fluid mud are analyzed by using both process-oriented and engineering-oriented methods. With the help of observation data and hydrodynamic and wave modeling, the relationship between the behavior of fluid mud and wave function, downslide of near-bed sediment, residual current regime and sediment availability are evaluated.

Chapter 4 analyzes the observations of ETM dynamics and saltwater intrusion in the DNC. ETM is a special phenomenon related fine sediment dynamics, study of its dynamics is essential to our understanding of the fine sediment characteristics in an estuarine surrounding.

Chapter 5 explores the dependency of settling velocity on SSC, salinity and temperature by an improved setting column. Determination of settling velocity has been regarded as a top priority in improving numerical modeling and conceptual understanding of fine sediment dynamics. So, the content is experimentally studied as an integral part of fine sediment dynamics.

Chapter 6 presents a sensitivity analysis on the factors governing the vertical structures of currents, salinity and SSC via a simplified 3D model, which is meant to evaluate the influences of those micro-scale effects (setting velocity, baroclinic, turbulence damping, and drag reduction) on ETM dynamics.

Chapter 7 models the seasonal variations of SSC regime in the North Passage by a real-case 3D Yangtze Estuary model. The numerical performances of multiscale seasonal control forces are assessed.

Chapter 8 forms the synthesis of the previous chapters, and the main findings and future perspectives related to estuarine fine sediment dynamics and estuarine waterway development are presented as our perspectives for future insights.

Hydrodynamic Processes[*]

Highlights

(1) Measured and modeled hydrodynamic processes in the Yangtze Estuary are presented.

(2) The water level rising along the main outlet is induced by the evolution of the whole river regime.

(3) The compressed estuarine environmental gradients have an indirect effect to the backfilling in the waterway. It strengthens the stratification effect near the area of estuarine turbidity maximum and enhances the tendency of up-estuary sediment transport.

[*]Parts of the chapter have been published in: **Wan, Y.**, F. Gu, H. Wu, and D. Roelvink (2014), Hydrodynamic evolutions at the Yangtze Estuary from 1998 to 2009, Applied Ocean Research, 47(0), 291-302, doi:10.1016/j.apor.2014.06.009.

2.1 Introduction

The Yangtze River (or Changjiang River) is the longest river in China and Asia, and the third-longest in the world, after the Nile in Africa and the Amazon in South America. The Yangtze River is historically, culturally, and economically important to China. The River is approximately 6300 km long and flows eastwards from its source in the Qinghai Province into the East China Sea at Shanghai. The Yangtze Estuary (*Figure 2-1*), three orders of bifurcations, a semi-enclosed coastal environment, is a free connection point for the river and the open sea, where salt and freshwaters meet and mix, and marine tides and riverine currents interact in the same location.

With the economic development along the eastern coastlines of China, especially in the Yangtze Delta area, sailing traffic has increased exponentially. To solve this problem, constructing a fairway, together with dredging strategies, to allow a deeper navigation channel is necessary. As a result, the conveyance of the channel will be enhanced and it will be able to meet the demand for more water depth. The Regulation Project of the Yangtze Estuary Deepwater Navigational Channel (DNC Project) was launched in 1998 after two years of investigation and study. The waterway depth was planned to be developed in three phases (see *Figure 2-2*) from 8.5 m in 2002 (Phase I) to 10 m in 2005 (Phase II) and to 12.5 m in 2010 (Phase III). The basic characteristic of these phases is the flow guidance contraptions becoming longer and narrower; as a result, the resistance of the path from the river to the sea is likely to increase irreversibly. Further, it should be noted that the result is enhanced due to both adjustment of the entire river regime and regional sea level rise.

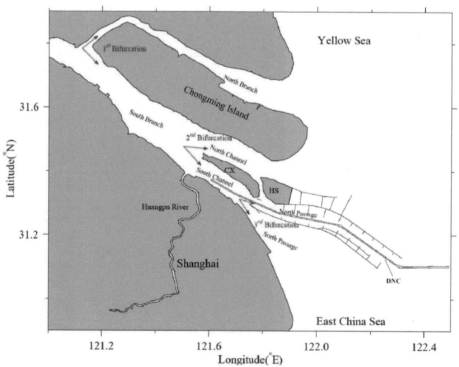

Figure 2-1. General layout of the Deepwater Navigational channel (DNC). The length of the DNC is approximately 90 km, and the elevation of the channel is maintained at -12.5 m by daily dredging (all the elevations, heights, and bathymetry in this thesis are referenced from the Lowest Astronomical Tide). CX and HS are two fixed boundary islands, Changxing Island and Hengsha Island.

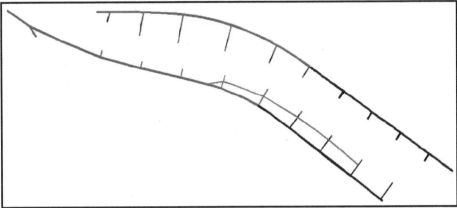

Figure 2-2.   The engineering progress of the DNC project (the pink lines indicate Phase I of the DNC project, the black lines indicate the extended engineering works of Phase II, and the red lines indicate the deployment of Phase III).

To evaluate the anthropogenic influences on the Yangtze Estuary over the last 12 years (1998-2009), the basically hydrodynamic parameters such as water level, tidal range, velocity, salinity, tidal prism, ebb dominance, bifurcation ratio, and suspended sediment concentration (SSC) are chosen to illustrate the spatiotemporal hydrodynamic evolutions along the main outlet, which is the main waterway of the Yangtze Estuary (shown by bold green lines, see *Figure 2-3*).

2.2   Methods and results

2.2.1   Measurement analysis

Because of the increase in sailing traffic in the Yangtze Estuary at the end of the last century, comprehensive and high-quality measurements and in situ field survey data could be collected, including basic hydrodynamic and bathymetric data. The positions of these measurement stations are shown in *Figure 2-3*, and the specified information about these measurements is shown in *Table 2-1*.

Table 2-1   Summary of the measurement data *(Wan and Qi, 2009)*.

Time	Classification	Wind condition (speed, direction)	River discharge (Datong station)	Sea water temperature
Feb. 10-16, 1998	Dry season, spring tide	4 m/s, E	22,000 m³/s	10 °C
Aug. 8-12, 2002	Wet season, spring tide	3 m/s, SE	49,000 m³/s	26 °C
Aug. 18-22, 2005	Wet season, spring tide	5 m/s, SE	40,000 m³/s	30 °C
Aug. 12-16, 2007	Wet season, spring tide	7 m/s, S (including a tropical storm)	50,000 m³/s	27 °C
Aug. 18-22, 2009	Wet season, spring tide	3 m/s, S	43,000 m³/s	30 °C

Multiscale physical processes of fine sediment in an estuary

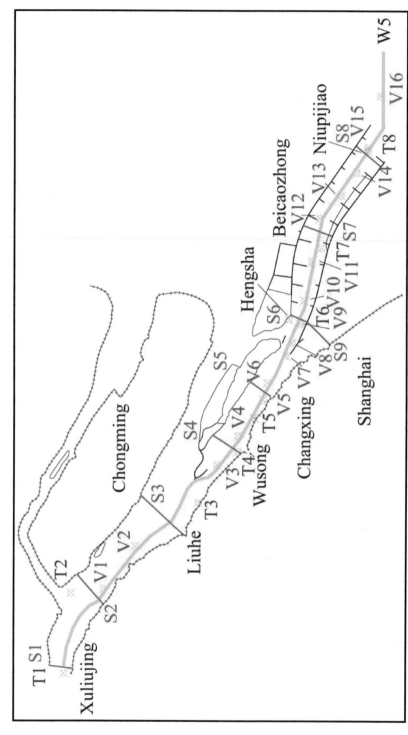

Figure 2-3. Deployment of the measurements (the blue points (T1-T8) are the tidal water level gauges ; the pink points (V1-V16) are the anchor stations for velocity, SSC, and salinity; the green bold line is the main outlet of the Yangtze Estuary; the red lines (S1-S9) across the river are the cross-section for calculating the tidal prism only in modeling).

As shown in *Table 2-1*, these measurements cannot be utilized for comparison, because the different meteorological conditions and the boundary conditions clearly vary. Despite this, some general and basic information can be obtained from *Figure 2-4* shown below.

(1) Although some measured data are unavailable (see *Tables 2-2* and *2-3* for data availability), the general trend observed from *Figures 2-4a* and *b* is that the higher high water level (HHWL) and lower low water level (LLWL) during spring tide period are increased by 0.4-1.2 m. However, the average annual sea level rise in this region is only 1-1.5 mm *(SOA, 2014)*, which means that the changes in HHWL and LLWL are not major contributors due to the sea level rise. The variation tendency of the HHWL and LLWL is clearly an overall change, which means that the variation is not primarily induced by regional engineering works. The overall water level rise in the entire Yangtze Estuary area should be triggered by the variation in the whole river regime.

(2) The ebb and flood tidal velocity (*Figures 2-4c* and *d*) changes appear more irregular and complex than the water level changes because of the different local hydrodynamic circumstances, in addition to data unavailability. Therefore, we consider the ebb dominance of the currents, which is a dimensionless index balancing the whole ebb tidal velocity and flood tidal velocity. Ebb dominance is obtained (as a percentage) by dividing the area of the ebb tide by the overall area (the area of the ebb tide plus the area of the flood tide) under the velocity-time (V-T) curve for a tidal cycle (*Figure 2-5*). Ebb dominance indicates the capacity of tidal currents to transport riverine sediments to the ocean. *Figure 2-4e* shows a evidence of tidal currents (or tidal energy) passing through the middle reach of the outlet (from Changxing to Niupijiao) and start vibrating, although the velocity increased only a little as the cross-section narrowed. This hydrodynamic condition may not favor a continuous displacement of sediment particles. Alternatively, similar information can be observed from *Figure 2-4f*. The ebb tide period increased from Hengsha to Beicaozhong, but decreased from Beicaozhong to Niupijiao.

(3) The SSC gradient (*Figure 2-4g*) at the reach from Wusong to Beicaozhong became steeper over time. The salinity gradient (*Figure 2-4h*) of the estuarine turbidity maximum (ETM) zone of the Yangtze Estuary *(Qi et al., 2010; Ma et al., 2011)*, at the reach from Hengsha to W5, clearly became steeper from 1998 to 2009.

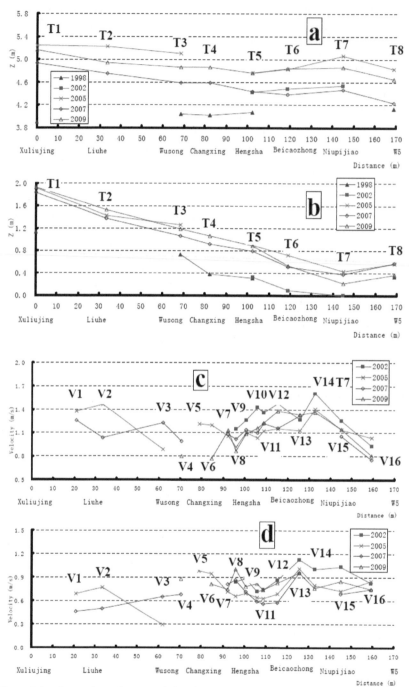

Figure 2-4. Comparison of measured data for different years along the main outlet of the Yangtze Estuary. (a) Higher high water level, (b) lower low water level, (c) tidal-period and depth averaged ebb tidal velocity, (d) tidal-period and depth averaged flood tidal velocity, (e) ebb dominance, (f) ratio of ebb period to flood period during a tidal cycle, (g) tidal period and depth averaged SSC, and (h) tidal period and depth averaged salinity).

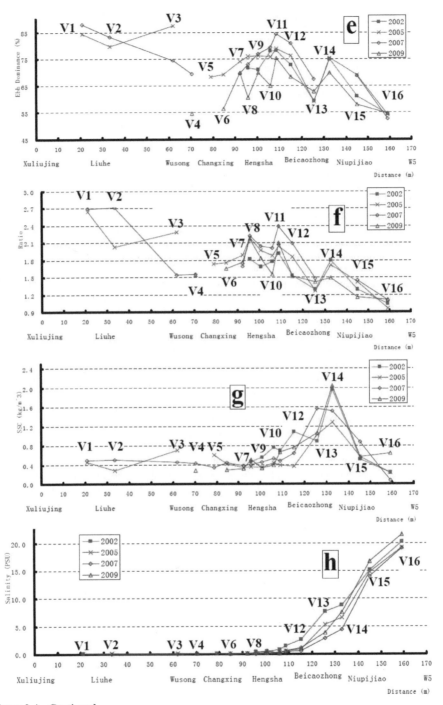

Figure 2-4   Continued.

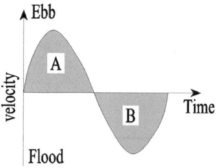

Figure 2-5. Sketch of the computation of ebb dominance (=A/(A+B)).

2.2.2 Numerical study

It can be argued that the above evidence, which was derived directly from measurement data, has a limitation imposed by the different meteorological and boundary conditions, resulting in unsatisfactory comparability of these data. Therefore in this study, we use numerical modeling to investigate the hydrodynamic evolution in the Yangtze Estuary from 1998 to 2009. In the numerical simulation, all of the conditions can be unified and adjusted easily. This enables the numerical method to overcome certain interferences and uncertainties more than the measurement analysis.

Due to the large-scale and complicated nature of the hydrodynamic circulation system of the Yangtze Estuary, the basic regime of the estuary is characterized as a three-stage bifurcation, four-river mouth split, shoal developed, available navigation channel alternated, mouth bar, and submerged delta stretching system. The tidal wave is strong enough to propagate to Datong, which is located approximately 650 km from Shanghai (see *Figure 2-6*), and its water and sediment frequently exchange with the Yellow Sea, East China sea, and Hangzhou Bay *(Shi, 2010)*. Further, the land boundary includes many islands and shoals, making it extremely complex, and the morphological variation and sediment characteristics in the estuarine and coastal system differ markedly throughout. All of these factors contribute to the motivation for developing an adaptable and flexible mathematical model of the Yangtze Estuary. Therefore, a two-dimensional, unstructured grid, hydrodynamic, and morphologic circulation model (named as SWEM, Shallow Water Equation Model) has been developed by our organization since 2003.

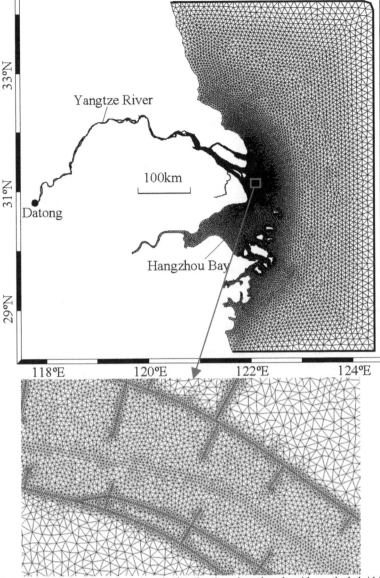

Figure 2-6.   Computational domain and mesh (the triangle and quadrangle grids are the hybrid fitting to local engineering constructions in *Figure 2-2*, with a fine resolution near the river plume; the dark dot and dark bold lines are the locations of the river and sea boundary, respectively).

2.2.2.1   Governing equations of SWEM

The two-dimensional shallow water equations that govern mass and momentum conservation are given below.

$$\frac{\partial \eta}{\partial t} + \nabla \cdot \vec{q} = 0 \tag{2-1}$$

$$\frac{\partial \bar{q}}{\partial t} + \nabla \cdot (\bar{q}\bar{V}) = -gD\nabla\eta - \frac{g\bar{q}|\bar{q}|}{C^2 D^2} - 2\bar{\Omega}\times\bar{q} + \nabla\cdot\left[A_H\left(\nabla\bar{q}+\nabla^T\bar{q}\right)\right]$$ (2-2)

where η is the surface elevation; $\bar{q}$ is the fluid flux vector; $\bar{V}$ is the fluid velocity vector; $D = H + \eta$ the water depth; g is the Earth's gravity; C is the Chezy roughness coefficient; and $\bar{\Omega}$ the rotation rate of the Earth. Further,

$$-2\bar{\Omega}\times\bar{q} \approx fq_y\bar{i} - fq_x\bar{j}$$ (2-3)

where f is the Coriolis parameter, $f = 2\Omega\sin\theta$, $\Omega = 7.2921e\text{-}5$, .. is the Latitude (rad). A_H is the horizontal eddy viscosity coefficient, which is obtained by the Smagorinsky sub-grid turbulence model. It is given by,

$$A_H = c_s\Delta^2\left[\left[\frac{\partial u}{\partial x}\right]^2 + 0.5\left[\frac{\partial v}{\partial x} + \frac{\partial u}{\partial y}\right]^2 + \left[\frac{\partial v}{\partial y}\right]^2\right]^{0.5}$$ (2-4)

where $c_s = 0.01 - 0.2$ is the Smagorinsky constant, and $\Delta = (da)^{1/2}$ is the area of the grid.

Following *Casulli and Walters (2000)*, a semi-implicit scheme is used to obtain an efficient numerical algorithm whose stability is independent of the surface gravity wave speed, wind stress, and bottom friction. The water surface elevation in the momentum equations and the velocity flux in the free surface equation are discretized by the Theta method, provided $0.5 \leq \theta \leq 1.0$ for stability. The wind stress and the bottom friction terms are discretized implicitly for stability. The advection terms in the momentum and scalar transport equations are incorporated into total derivatives, and solved using a total variation diminishing (TVD) scheme to avoid the Courant number constraint. In addition, to remove the orthogonality constraint of the grid, most terms in momentum equations are evaluated at the center of the cell using a finite volume scheme, and then interpolated onto the face conservatively. The details of the above procedures are described elsewhere *(Qi et al., 2010; Ma et al., 2011)*.

2.2.2.2 Model setup

(1) Computation domain and mesh
    To accurately model the tidal water interchange and tide propagation of the entire Yangtze Estuary, the computation domain (*Figure 2-6*) is set to approximately 650 km×680 km, covering the entire Hangzhou Bay, part of the East China Sea and the Yellow Sea, which is the entire tidal area of the Yangtze River.
    The mesh of the domain is as follows: the studied areas such as the North Passage, dikes, groins, and nearby shallow areas are all refined and shape adapted. The total number of nodes is 69,482, and the number of cells is 136,084.
(2) Open ocean tidal boundaries
    The water level is described as an astronomic tidal boundary condition by specifying the tidal components (see *Figure 2-6*). The tidal components were derived from the NaoTide data set *(Matsumoto et al., 2000)* (*www.miz.nao.ac.jp/*). During the modeling, 16 tidal constituents (M2, S2, N2, K2, K1, O1, P1, Q1, MU2, NU2, T2, L2, 2N2, J1, M1, and OO1) are selected with the annual mean sea water level distribution along the open boundary.
(3) River boundary
    To simulate the tidal propagation and make it fully dissipated along the river, the Datong station, which is the tidal limit of the Yangtze Estuary *(Qi et al., 2010)* (see *Figure 2-6*), is taken as the river boundary. The daily river discharge at the Datong can be obtained from *http://yu-zhu.vicp.net*.
(4) Simulation configuration
    Five scenarios (1998, 2002, 2005, 2007, and 2009) are modeled to simulate the hydrodynamic development of the Yangtze Estuary. In each scenario, the bottom boundary is

updated by the measured bathymetry. The engineering progress of the DNC project is also depicted by adjusting the local topography according to the actual elevation of the engineering structures. The five scenario simulations are forced by unified boundary conditions. The unified boundary conditions are as follows: (i) the Datong discharge is 40,000 m³/s, (ii) the tidal boundary and bottom roughness are the same, (iii) wind, wave, sediment transport, saltwater intrusion, and morphology changes are not included in the hydrodynamic modeling system, and (iv) the simulation period is from August 12-16, 2007 (wet season, spring tide), and the time step interval of the simulation is 3 s.

2.2.2.3 Validation

As an essential procedure of the model setup, the validation and calibration of the SWEM are frequently carried out when new measurement data are available *(Qi, 2007; Qi et al., 2010)*. The river flow measurement and astronomic tidal levels are selected as the boundary forcing conditions of the model. The measured data (water levels and depth-averaged currents) in the five real-world cases in *Table 2-1* are all calibrated against the corresponding modeled data. *Figures 2-7* and *8* present a reasonable agreement between the observed and modeled water levels and depth-averaged currents, respectively. Due to the space limitation of this chapter, other specified comparisons between the observation and simulation are presented in *Qi (2007)*.

Recently, the model skill score (S, see *equation 2-5*) has been commonly used to evaluate numerical modeling performance *(Willmott et al., 1985; Li et al., 2005; Ma et al., 2011; Wu et al., 2011)*. The model skill scores of different calibrations are calculated in *Tables 2-2* and *2-3*. The scores of water level range 0.90-0.98 and the scores of currents range 0.71-0.95. The performance of the calibration indicates that the model showed a reasonable capability.

$$S = 1 - \frac{\sum_{i=1}^{N}\left|X_{mod} - X_{obs}\right|^2}{\sum_{i=1}^{N}\left(\left|X_{mod} - \overline{X}_{obs}\right| + \left|X_{obs} - \overline{X}_{obs}\right|\right)^2} \tag{2-5}$$

where X is the variable of interest, $\overline{X}$ is the time mean, and X_{mod} and X_{obs} are the modeled and observed results, respectively. A Larger S value denotes a better agreement between the observation and simulation; $S = 1.0$ means perfect modeling performance level.

Table 2-2. Model skill scores of water levels for each calibration.

Station	T1	T2	T3	T4	T5	T6	T7	T8
1998	-	-	0.93	0.93	0.91	-	-	-
2002	-	-	-	-	0.94	0.92	0.95	-
2005	0.94	0.96	0.93	-	0.94	0.97	0.98	0.95
2007	0.95	0.96	0.92	0.93	0.95	0.97	0.99	0.97
2009	0.97	0.92	0.95	0.94	0.96	0.93	0.97	0.97

Note that "-" means no measured data available.

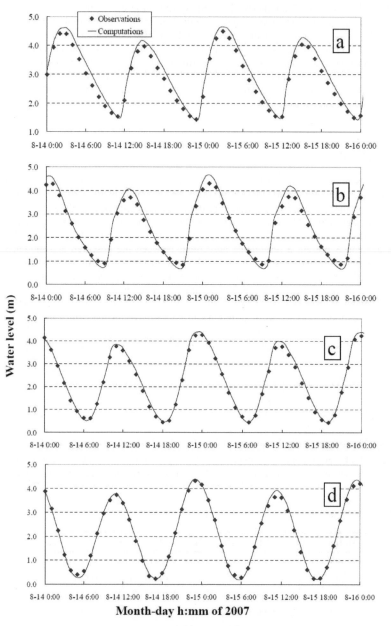

Figure 2-7. Comparisons of observed and simulated water levels at stations T1 (a), T4 (b), T7 (c), and T8 (d) during August 14-16, 2007.

Table 2-3.   Model skill scores of tidal currents for each validation.

Stations	V1	V2	V3	V4	V5	V6	V7	V8	V9	V10	V11	V12	V13	V14	V15	V16
1998	-	-	-	-	-	-	-	0.89	0.80	0.77	0.89	0.79	0.82	0.94	0.83	0.85
2002	-	-	-	-	-	-	-	0.87	0.83	0.85	0.81	0.86	0.84	0.90	0.78	-
2005	0.84	0.79	0.82	-	0.79	0.84	0.83	0.73	0.90	0.89	0.80	0.92	-	0.85	0.71	-
2007	0.92	0.76	0.81	0.88	-	-	0.76	0.89	0.95	0.79	0.84	0.89	0.85	0.76	0.84	0.82
2009	-	-	-	0.80	-	0.78	0.84	0.92	-	-	-	-	-	-	-	-

Note that "-" means no measured data available.

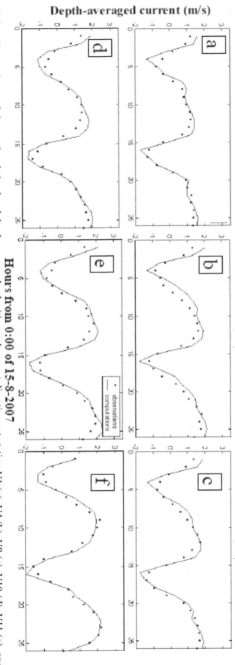

Depth-averaged current (m/s)

Hours from 0:00 of 15-8-2007

Legend: • observations — computations

Figure 2-8.   Comparisons of observed and simulated depth-averaged velocities and current directions at station V1 (a), V4 (b), V8 (c), V10 (d), V11 (e), and V13 (f) during August 15-16, 2007.

2.2.2.4   Results

In the numerical simulation, the same parameters used in the measurement analysis are chosen to illustrate the evolution of hydrodynamics in the Yangtze Estuary. The results are presented in *Figure 2-9*, from which certain clear and significant observations can be made, as described below.

(1) HHWL and LLWL increased approximately 0.2-0.5 m gradually from Xuliujing to W5 during the past decade (*Figures 2-9a* and *b*). This increase may not only be affected by the DNC project because the engineering deployment of the project is located only between Hengsha and Niupijiao, but the water level is affected along the whole reach of the DNC project. This implies that the evolution of the overall river regime (*Figure 2-10*) is the main factor for the water level rise in the Yangtze Estuary. As *Figure 2-10* shows, there is more sedimentation than erosion in this region, so the total volume of the Yangtze Estuary riverbed is reduced, increasing the water level. Moreover, the range of LLWL variation at Hengsha, which is the entrance to the DNC, is greater than that at any other reach, especially before and after Phase II of the DNC project. This change is clearly induced by the project, and it makes the gradient of the water surface during ebb times from Hengsha to Niupijiao much steeper than before.

(2) The capacity of erosion in the North Passage is enhanced, which is considered to be one of the main regulation purposes of the DNC project, and the ebb tidal velocity increased sharply from Hengsha to Niupijiao (*Figure 2-9c*) through the path of narrowing cross-section. As a result, after 2002 (Phase I of the DNC project), the resistance for flood currents increased and the flood tidal velocity (*Figure 2-9d*) decreased, especially near Niupijiao.

(3) The greatest change in the tidal prism along the DNC is at the North Passage (S6 and S7) and the ebb and flood tidal prism decreased simultaneously (*Figures 2-9e* and *f*). The flood tidal prism decreased sharply, especially at Beicaozhong, where the sedimentation and back-silting are the highest along the channel *(Liu et al., 2011)*.

(4) The ebb dominance of the currents (*Figure 2-9g*) increased on the whole along the DNC, but at the last reach of the outlet, it decreased sharply from 85% to 55% along only a 20 km spread.

(5) The flow portion (*Figure 2-9h*) for the North Passage, the 3rd bifurcation of the Yangtze Estuary, experienced a gradual decrease.

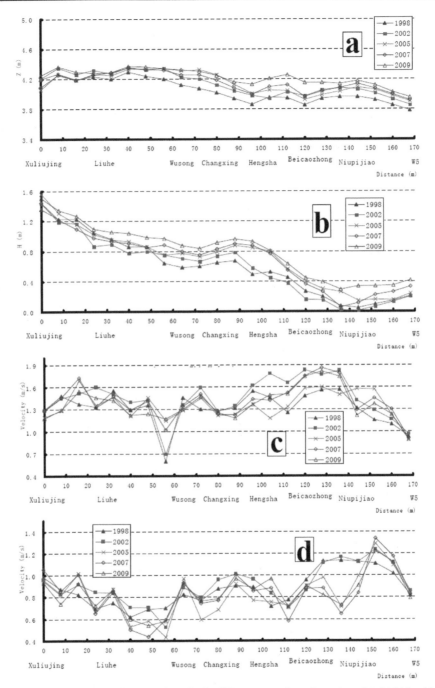

Figure 2-9. Comparisons of simulated results for different years along the main outlet. (a) higher high water level, (b) lower low water level, (c) depth averaged ebb tidal velocity, (d) depth averaged flood tidal velocity, (e) ebb tidal prism per day, (f) flood tidal prism per day, (g) ebb dominance, and (h) ratio of flow portion at the 3rd Bifurcation). The location of the cross sections for the tidal prism can be found in *Figure 2-3*. The position of these numerical observation stations is along the main outlet, and the distance of each two neighboring points is 10 km.

31

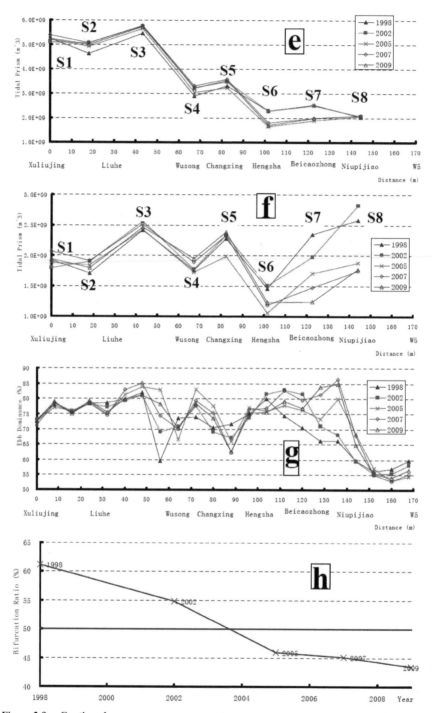

Figure 2-9. Continued

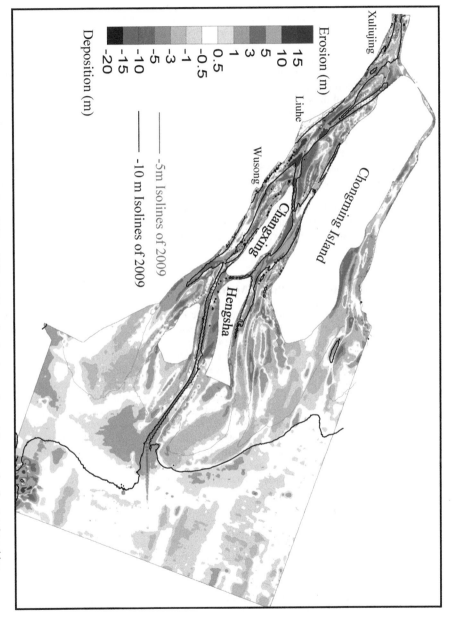

Erosion (m)

15
10
5
3
1
0.5
-0.5
-1
-3
-5
-10
-15
-20

Deposition (m)

——— -5m Isolines of 2009

——— -10 m Isolines of 2009

Figure 2-10. Riverbed changes in the Yangtze Estuary from 1998 to 2009; the blue color denotes erosion, and the red color denotes deposition.

2.3 Discussion

Using a field and numerical investigation, this study seeks to increase our understanding of the recent hydrodynamic evolution of the Yangtze Estuary, especially analyzing the impact of the DNC project. Furthermore, its potential influence on sediment transport and back-silting within the navigational channel is discussed. Finally, a conceptual model for the evolution is proposed.

According to the water level variations (*Figures 2-4a, b* and *2-9a, b*), it can be observed that the water level along the main outlet of the Yangtze Estuary increased from 1998 to 2009. This is a long-term change and is not due to a locally adjusted engineering project. This means that the variation of the whole river regime (including the natural geomorphodynamic processes and local topography feedback from extreme metrological events and artificial engineering interventions) at the Yangtze Estuary contributes to the water level rise. However, from *Figure 2-9h*, the flow portion ratio at the 3rd Bifurcation was decreasing during the engineering progress of the DNC project. The dramatic change is only coincidental with the deepening of the navigational channel from 1998 to 2009. This indicates that the decrease in the flow portion ratio at the 3rd bifurcation might be due to the DNC project and the corresponding bathymetry transformation in the North Passage. The above change in the North Passage is only one example of the contributing factors for the whole river regime variation of the Yangtze Estuary.

Further, more hydrodynamic evolutions compress the estuarine environmental gradient (such as the salinity and SSC), which pose at least two underlying problems, as described below.

2.3.1 Stratification near the ETM area

From *Figure 2-4g*, the SSC gradient near the ETM area (from Wusong to Beicaozhong) is becoming steeper from 1998 to 2009. The compressed along-channel SSC gradient would favor fine sediment particle aggregation near V14 (see *Figure 2-4g*), which is always the location of the ETM area, regardless of the fact that the hydrodynamic and morphological conditions have extensively changed the entire estuary. This means that the balancing position of tidal energy and riverine flow is almost constant, but the amplitude of the ETM is decreasing. This would lead to fluctuating suspended sediment aggregation that is trapped near center of the ETM. Because of the nature of the cohesive sediment, it prefers to flocculate, accelerating the settling velocity and *(Winterwerp, 2002; Maggi, 2007)* sustaining high concentrations near the bottom. This begins to create a deformation of the SSC vertical profile, referred to as the SSC stratification effect. It will further reduce the turbulence *(Ivey et al., 2008)* and alter the vertical profile of the currents *(Burchard and Baumert, 1998)*. It should be noted that the stratification effect will be enhanced in response to channel deepening *(Winterwerp, 2011a)*.

2.3.2 Up-estuary sediment transport

The effects of the barotropic and baroclinic pressure gradients on the water motion are implicit in the momentum balance equation *(de Nijs, 2012)*, so their direct contribution to the structure of currents is not straightforward. To determine the dominant term, the accelerations induced by the barotropic and baroclinic pressure gradients are quantified for estimation of their influence on the fluid motion.

In the horizontal momentum equation, the pressure gradient term consists of the barotropic and baroclinic pressure gradient terms.

$$\frac{1}{\rho_0}\frac{\partial p}{\partial x} = g\frac{\partial \xi}{\partial x} + \frac{g}{\rho_0}[\int_z^\xi \frac{\partial \rho}{\partial x}dz'] \tag{2-6}$$

where x is the along-channel direction, p is the fluid pressure (including the hydrostatic (barotropic) pressure and the density gradient induced (baroclinic) pressure), ρ_0 is the reference density, ξ is the water level, ρ is the fluid density, and g is the gravitational acceleration.

As shown in *Figure 2-4g* and *h*, the horizontal gradient term increases directly, and the vertical density gradient decreases indirectly with the stratification effect, and the direction of the baroclinic pressure gradient is always landward, this term is the dominant control force for material advection near the ETM area *(de Nijs and Pietrzak, 2012)*. However, during flooding and ebbing *(Figures 2-9a* and *b)*, several inverted slopes exist in the water surface curves for near the lower reach of the North Passage, meaning that the direction of barotropic pressure gradient is sometimes up-estuary also. Then, the up-estuary sediment transport is enhanced near the place of inverted water level slope. In this situation, the process of delivering riverine sediment to the ocean is hampered near the ETM area, which should result in maintenance issues in the navigational channel.

Over the past decade, slow and remarkable changes accumulated, with clear impact on the Yangtze Estuary, but some aspects of the variation may not have been observed yet. Most importantly, this study illustrates the changes from both field and numerical perspectives. The future impacts by the hydrodynamic evolution are uncertain, with unknown influence on the well-being of the Yangtze Delta area. It is certain that we need a healthy, functioning estuary for sustainable development *(Saeijs, 2008)*. It is well known that estuaries and coasts are amongst the most important and valuable environments in the world. The Yangtze Estuary provides high productivity, attractive and active habitats, and rich biodiversity. However, the coastal ecosystem is very vulnerable; therefore, it needs greater attention and more innovations to actively guide the future trends of hydrodynamic evolution in the Yangtze Estuary. This study is designed to cast a new light on the development of the entire Yangtze Estuary.

2.4   Conclusion

By contrasting and analyzing the field survey and numerical simulation results of the past decade (1998-2009), the following conclusions can be drawn.

(1) HHWL and LLWL increased approximately 0.2-0.5 m from Xuliujing to W5. The water level rise is induced by the variation in the whole river regime, including natural morphodynamic processes and local topography feedback from extreme metrological events and artificial engineering interventions.

(2) The decrease in the flow portion ratio at the 3rd Bifurcation is directly due to the DNC project and the corresponding bathymetry adjustment in the North Passage.

(3) The estuarine environmental gradients (the salinity and SSC) have been compressed, and the fresh-salt gradient became steeper according to the field data. This has had an indirect effect (strengthening the stratification effect near the ETM area and enhancing the up-estuary sediment transport) on the waterway back-silting, as mentioned in the literature *(Chen and Lin, 2000)*.

Fluid mud dynamics[*]

Highlights

(1) A storm-induced fluid mud event is investigated in a muddy-estuarine navigational channel from the views of both process-oriented and engineering-oriented approaches.

(2) Both suspended particulate matter availability and local residual flow regime are of critical importance to the trapping probability of sediment and the occurrence of fluid mud.

(3) The fluid mud dynamic process is both an advective and a tidal energy influenced phenomenon.

[*] Parts of the chapter have been published in: **Wan, Y.**, D. Roelvink, W. Li, D. Qi, and F. Gu (2014), Observation and modeling of the storm-induced fluid mud dynamics in a muddy-estuarine navigational channel, Geomorphology, 217(0), 23-36, doi:10.1016/j.geomorph.2014.03.050.

3.1 Introduction

The occurrence of fluid mud is widely covered and commonly witnessed in many locations, such as estuaries *(e.g. Mehta, 1989; Winterwerp, 1999)*, lakes *(e.g. Li and Mehta, 2000; Bachmann et al., 2005)*, rivers *(e.g. Wang, 2010)*, waterways *(e.g. Li et al., 2004)* and even open sea *(e.g. Puig et al., 2004)*. Fluid mud exists in the water column *(Figure 3-1* and *Table 3-1)* as a transitional stage *(McAnally et al., 2007a)*, when the net rate of sediment falling from the upper suspension layer into the bottom layer exceeds the dewatering rate of the high-concentration sediment-water mixture, and the bonds of the interconnected matrix structure are not strong enough to form an erosion-resisting consolidated layer. The characteristics of fluid mud differ significantly from those of both suspensions above and the consolidated bed below. The temporal transition status varies quickly in response to sediment availability and intensity of currents (when fluid mud is "left alone" it will consolidate).

Consider the condition of sediment availability or supply, it may relate to micro-scale sediment mixing, such as flocculation and hindered settling *(Le Hir et al., 2000)*; it could form a stepped vertical profile of suspended sediment concentration (SSC) and trap sediment in the near-bed layer. At the same time, the sediment supply is also associated with macro-scale sediment movement and circulation *(Shi, 2010)*, where transport of enough fine sediment mass from nearby shoals and beaches to the navigational channel favours the formation of a fluid mud layer. The current dynamics can also be divided into micro- and macro-scale processes, where the micro-scale processes include flocculation settling, turbulence damping, drag reduction and some stratification effects of flow, while the macro-scale refers to the regime of currents, residual circulation, and tidal asymmetry and so on.

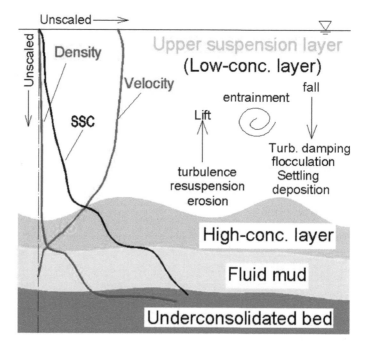

Figure 3-1. Perspectives on the dynamics of the fluid mud in the water column (SSC is suspended sediment concentration)

Table 3-1.   Classification of sediment-water mixtures (suspensions)

Classification	Alias	Bulk density [a] (kg/m³)	SSC [a] (kg/m³)	Motion	Fluid behavior
Low-concentrated layer	dilute suspensions *(Bruens, 2003)*	1000-1020	<10	movable	Newtonian
High-concentrated layer	concentrated benthic suspensions (CBS) *(Winterwerp et al., 2002)*	1020-1050	1-several 10s	movable	Newtonian
Fluid mud	hyperconcentrated benthic layer *(Winterwerp, 1999)*; lutoclines *(Kirby and Parker, 1983)*	1050-1300	several 10s-a few 100s	movable, erodible, depositable	Bingham
Consolidated bed	settled bed *(Manning et al., 2010)*	>1300	>750	erodible, depositable	Solid-state

Note: (a) the ranges of bulk density and SSC are varied by site-specific situations, and are highly dependent on local sediment compositions and hydrodynamics

Therefore, there are two types of viewpoint from which to study the dynamics of fluid mud. The first approach is process-oriented or micro-mechanism driven, which is conducted primarily by sedimentologists, geomorphologists and oceanographers; they focus on some responses and influences on sedimentary processes and vertical profiles of currents and SSC, such as flocculation, re-suspension, deposition, erosion, turbulence damping, drag reduction, density flow, and turbidity maximum. The second method can be called engineering-oriented or macro-mechanism driven, which is the approach chosen mostly by hydraulic and coastal engineers, who are concerned with the horizontal and overall regime of currents and sediment; typically the keywords in this type of research are residual circulation, tidal asymmetry, sediment availability, flow regime and so on.

Many complex physical processes are related to the formation of fluid mud, such as flocculation, settling and mixing, deposition, re-entrainment, gelling, consolidation, liquefaction and erosion *(Winterwerp, 1999)*. In addition, *McAnally et al. (2007b)* showed that physics, such as rheology, as well as chemical oceanography and microbiology also play a large role in fluid mud behavior. Many efforts *(e.g. Wolanski et al., 1988; de Wit, 1995; Kineke et al., 1995; Ali et al., 1997; Shi, 1998; Le Hir and Cayocca, 2002; Vinzon and Mehta, 2003; Guan et al., 2005; Winterwerp, 2006; Hsu et al., 2007; McAnally et al., 2007a)* have been dedicated to investigating the formation of fluid mud. Among those studies, the effect of wave and storm processes on fluid mud has attracted considerable attention in recent years. *McAnally et al. (2007a)* pointed out that fluidization of soft sediment beds by wave agitation is one of three principal mechanisms of the fluid mud formation. *Li et al. (2004)* suggested that the formation of fluid mud phenomena may fall into three categories: slack water, storm and salt wedge. Through measurement data from two moored tripods, *Traykovski et al. (2000)* showed that the fluid mud could be trapped within the wave bottom boundary layer. Based on tripod data, *Puig et al. (2004)* also provided a clear picture of the influence of surface-wave activity on the rapid generation of a sediment gravity flow (fluid mud) by development of excess pore water pressure during storms. *Warner et al. (2008)* utilized ROMS and SWAN models to reveal that bottom sediment resuspension is controlled predominantly by storm-induced surface waves and transported by the tidal- and wind-driven circulation. With the aid of laboratory experiment, *van Kessel and Kranenburg (1998)* showed that the wave-induced liquefaction (fluid mud) of subaqueous mud layers may be a mechanism of rapid sedimentation observed in navigational channels after storms. In summary, wave energy has the potential to resuspend, release and load sediments in a submarine layer, which facilitates the formation of fluid mud; in particular, if it can stir up large quantities of sediment over mudflats and in shallow areas under wind wave conditions. In short, storm generation is considered one of the most significant causes of fluid mud.

In this chapter, firstly, a storm-induced fluid mud event in a muddy-estuarine navigational channel is studied. Secondly, wind wave propagation is modeled to examine the condition of sediment availability under a cold-air front, and three-dimensional (3D) hydrodynamics are simulated to achieve a better understanding of the major mechanism determining the dynamics of fluid mud. Finally, both process-oriented and engineering-oriented methods are employed to investigate the possible factors influencing the mechanisms and transport processes of this storm-induced fluid mud event.

3.2   The fluid mud event

Recently, a 90 km-long and 12.5 m-deep (all elevations, heights and water depths in this chapter are referred to the Lowest Astronomical Tide) deepwater navigational channel (DNC) in the North Passage of Yangtze Estuary, China was completed. The engineering construction of the DNC was launched in 1998, the water depth of the navigational channel was developed in three steps from 8.5 m in 2002 (Phase I), 10 m in 2005 (Phase II), and to 12.5 m in 2010 (Phase III) (see *Figure 3-2*).

In recent years, according to in situ measurement data, the back-silting and siltation along DNC is so severe that it requires many efforts to study the possible mechanisms behind the massive trapping of fine cohesive suspended sediments in the river- and tide-dominated estuary. One specific case, a fluid mud event, may illustrate the fine sediment dynamics and trapping processes clearer than regular conditions.

At the end October of 2010, an intensified cold-air front affected most areas of southeastern China. The observed thickness of the fluid mud resulting from this storm along the channel was about 1-5 m. The field investigation of this event was expected to evaluate and analyze the distribution, influence and dynamics of fluid mud. First of all the forcing conditions of the fluid mud event are introduced, and then the measurement data are investigated to show the dynamics of the fluid mud.

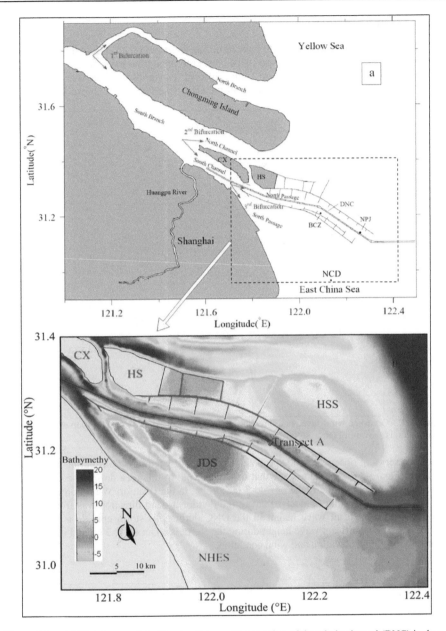

Figure 3-2. (a) Sketch map of the Yangtze Estuary. The location of the whole channel (DNC) is shown in the upper figure; the length of the DNC is about 90 km. CX and HS are two fixed boundary islands, Changxing Island and Hengsha Island; NCD is the site of wind and wave data shown in *Figures 3-3, 3-4* and *3-6*. BCZ and NPJ are the tidal gauging stations. (b) Bathymetry with constructed dikes and groins in the North Passage of 2010 (JDS, HSS, and NHES denote Jiuduansha Shoal, Hengsha Shoal and Nanhui East Shoal respectively), the location of the transect shown in *Figure 3-8* is Transect A. Phase I of the DNC project is shown in pink lines, Phase II is shown in black, and Phase III is in red.

3.2.1 Forcing conditions

This cold-air front was reported to be recorded as the strongest cold front during the past 5 years in China; the temperature dropped 6-14 °C sharply along the track of this winter storm[1]. The coastlines of eastern and southern China suffered a long-lasting intensive wind-induced storm, the maximum wave height observed in the DNC (the location is shown in *Figure 3-2*) was up to 4.35 m.

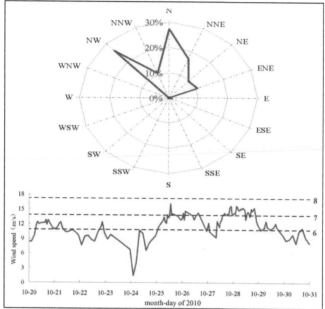

Figure 3-3. Observed wind speed and direction during the storm at NCD (the location of the wind and wave station is shown in Figure 3-2-a)

Time series of observed wind speed and frequency of wind directions during the storm are shown in *Figure 3-3*. The data show that maritime wind in this area during the cold-air front were estimated to be in the range of 6 to 7 (according to the Beaufort Wind Force Scale), and the wind was long-lasting (almost 10 days); the dominant wind direction was from the north. The observed significant wave height in the DNC during this storm episode is presented in *Figure 3-4*. The significant wave height during the storm is about 0.5-2.2 m, and its variation is related to local tidal elevation and wind condition.

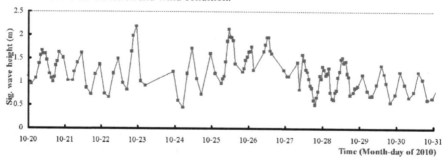

Figure 3-4. Time series of observed significant wave height during the storm at NCD (the water depth at NCD is about 9±2 m, the significant wave period is about 4-7 s; some of the hourly or three-hourly data are missing due to signal dropout.)

---

[1] Source: ChinaNews, *www.chinanews.com/gn/2010/10-25/2611092.shtml* (in Chinese)

3.2.2   Dynamics of the fluid mud

From 18th Oct. to 28th Nov. of 2010, 13 profiles of along-channel fluid mud field investigations were collected. The observed channel-longitudinal dynamics of fluid mud (from echosounder[2] data) along the DNC just after the storm is shown in *Figure 3-5*, the time-series of the total volume of observed fluid mud in the channel is given in *Figure 3-6*, typical along-channel and cross-channel density profiles of the fluid mud are also given in *Figures 3-7* and *3-8*. Several important characteristics could be drawn out from the figures.

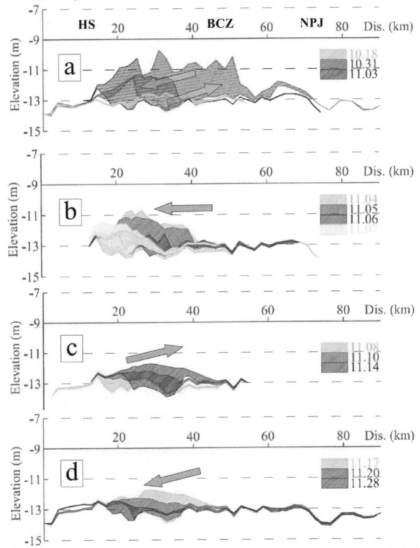

Figure 3-5.   Time-series of observed along-channel dynamics of fluid mud (by echosounders) along DNC just after the storm (location of the transect is shown in *Figure 3-2*, and this transect is in the middle of the whole DNC channel). (a) Tidal energy firstly decreased from 18-Oct. to 31-Oct., then it increased; (b) tidal energy was increasing from 04 to 07 Nov.; (c) tidal energy was decreasing; (d) tidal energy was increasing.

---

[2] The low and high frequency of echosounder applied in this study was 33K Hz, and 210K Hz, respectively.

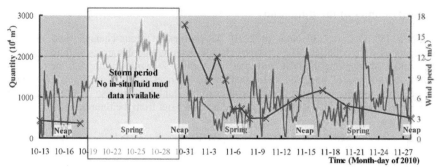

Figure 3-6. Time series of observed total volume of fluid mud in the whole channel (the pink line is the observed wind speed at NCD, the beginning and end of the whole channel (DNC) are shown in Figure 3-2-a).

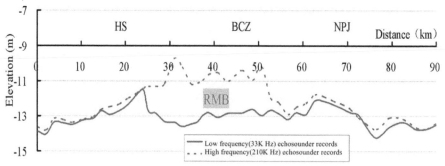

Figure 3-7. Typical along-channel density profile of the fluid mud (this measurement was collected on 10-31-2010; RMB is the abbreviation of reconstructed temporary mouth bar).

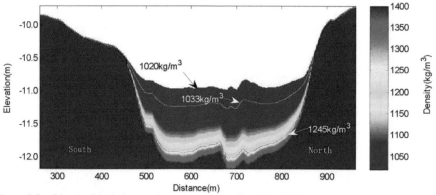

Figure 3-8. Measured typical cross-channel density profile of the fluid mud, the location of this transect is in the middle reach of the North Passage, which is shown in *Figure 3-2b*. The values 1020, 1033 and 1245 kg/m³ are the density of the near-bed suspensions, initial fluid mud layer and weakly consolidated bed; the cross-channel density profile is measured by the STEMA DensiTune and the measurements were made on 10-31-2010.

(1) Overall, the fluid mud during this time is mainly controlled by the intensity of the wind. It seems that a temporary mouth bar was constructed (*Figure 3-7*) by fluid mud according to the field data from the echosounders, and the position of this storm-induced fluid mud is more varied in the upper reach of the North Passage (from HS to BCZ).

(2) The duration of the fluid mud event lasted for quite a long period after the storm

(*Figure 3-6*), despite some efforts by dredgers to drive away fluid mud and resuspend the sediment particles (avoiding their consolidation) that were not totally consolidated.

(3) Fluid mud was only observed along the dredged channel (*Figure 3-8*) and barely appeared over the nearby shallow shoals.

(4) The oscillation of fluid mud (as shown by the orange arrows in *Figure 3-5*) shows an obvious coincidence with the tidal variations. As shown in *Figures 3-5a, b* and *d*, when the tidal energy increases (i.e. in *Figure 3-5b*, the tides transfer from neap tide to spring tide), the thickness and total volume of fluid mud decrease, and the horizontal central position of fluid mud layer moves up-channel. Generally, if the tidal range increases within the estuary, the mixing zone between tides and riverine flow will move upstream *(Buschman et al., 2009)*. The fluid mud layer also moves with the mixing zone. It means that with higher tidal energy, fluid mud is likely to be dispersed and washed away. In contrast, as shown in *Figures 3-5a* and *c*, when the intensity of tidal mixing decreases, the thickness and quantity of fluid mud increase and the central position moves down-channel. With lower tidal energy, fluid mud is likely to accumulate and enlarge. So, in short, it indicates that the transport of fluid mud is an advective phenomenon determining the central position of fluid mud layer along the channel, and also is a tidal energy-influenced phenomenon controlling the accumulation and erosion of fluid mud.

3.3   Modeling

We know that sediment is mainly agitated by wave effects and transported by tidal currents during some extreme wave weather events, for instance storm surge and typhoon *(Kineke et al., 1995; Dalrymple et al., 2008)*. So here numerical evidence for the availability of suspended particulate matter (SPM) is provided by wave modeling for examining the sediment source of the fluid mud event. At the same time, the net transport of tidal currents and sediment transport pathway is investigated by 3D hydrodynamic modeling for hindcasting the transport processes of fluid mud.

3.3.1   Wave propagation modeling

SWAN *(Booij et al., 1999; Zijlema, 2010)* (v40.81, based on an unstructured grid) is a third-generation spectral wave model that computes random, short-crested wind-generated waves in coastal regions and inland waters. Physical processes represented are wind input, propagation, wave interactions, energy dissipation, wave-breaking, refraction, reflection, whitecapping, diffraction, wave-induced set-up and transmission around obstacles.

Wave propagation for the fluid mud event is simulated by SWAN coupled with hydrodynamic model during the storm period in the entire estuarine area. The wave model is calibrated against significant wave heights in the DNC (see *Figure 3-9* for the locations of the observations); reasonable agreement between observation and simulation is presented in *Figure 3-10*. *Figure 3-11* shows the simulated significant wave heights and wave directions at the peak of the storm.

As *Jing and Ridd (1996)* stated, for sediment transport problems, co-action between waves and currents has two significant effects: (1) it results in a change in eddy viscosity coefficient, and (2) it enhances the bed turbulence and causes an increase in bottom shear stress. In *Section 3.4.2*, wave-induced effect on bottom stress will be presented.

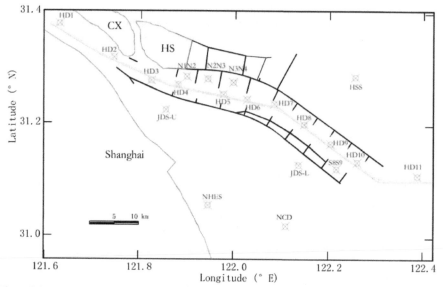

Figure 3-9. Location of the mooring stations for SSC and salinity.

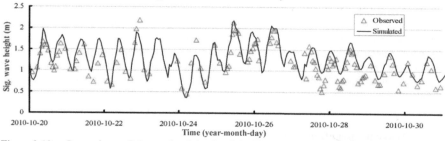

Figure 3-10. Comparisons of observed and simulated significant wave heights at NCD.

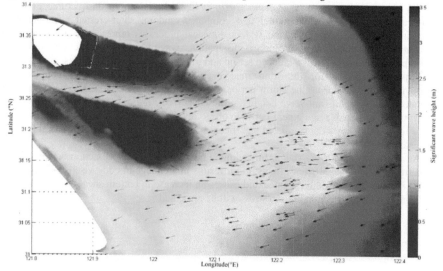

Figure 3-11. Simulated significant wave heights and directions at the peak of the storm (Time: 25-10-2010, 10:00).

3.3.2 3D hydrodynamics simulation

The Yangtze Estuary is unlike those simple funnel-shaped or delta-shaped estuaries. Its huge size, its extremely complex tidal hydrodynamic circulation system with three level bifurcations (see *Figure 3-2*), and the dynamics of cohesive fine sediment present great challenges in understanding sediment transport in relation to maintenance of the navigational channel. The median grain size of suspended sediment in the muddy estuary is about 4-11 μm *(Wang and He, 2007; Guo and He, 2011)*, and the bed load is 8-120 μm *(Li et al., 2008; Hu et al., 2009b)*.

A three-dimensional, generalized σ-coordinate system, unstructured finite volume grid, baroclinic estuarine circulation model was built to simulate the complex hydrodynamics in the estuary. The governing equations of this model in a sigma coordinate system (x,y,σ,t) could be read as follows,

$$\frac{\partial u}{\partial t}+u\frac{\partial u}{\partial x}+v\frac{\partial u}{\partial y}+\frac{\omega}{H}\frac{\partial u}{\partial\sigma}=-\frac{1}{\rho_0}\frac{\partial p_n}{\partial x}+\frac{\partial}{\partial x}(2A_h\frac{\partial u}{\partial x})+\frac{\partial}{\partial y}[A_h(\frac{\partial u}{\partial y}+\frac{\partial v}{\partial x})]+\frac{1}{H}\frac{\partial}{\partial\sigma}(A_v\frac{1}{H}\frac{\partial u}{\partial\sigma})$$
$$-g\frac{\partial\xi}{\partial x}-\frac{gH}{\rho_0}\frac{\partial}{\partial x}[\int_\sigma^0\Delta\rho d\sigma']+\frac{g}{\rho_0}\frac{\partial H}{\partial x}[\int_\sigma^0\sigma'\frac{\partial\Delta\rho}{\partial\sigma'}d\sigma']+fv \tag{3-1}$$

$$\frac{\partial v}{\partial t}+u\frac{\partial v}{\partial x}+v\frac{\partial v}{\partial y}+\frac{\omega}{H}\frac{\partial v}{\partial\sigma}=-\frac{1}{\rho_0}\frac{\partial p_n}{\partial y}+\frac{\partial}{\partial x}(2A_h\frac{\partial v}{\partial y})+\frac{\partial}{\partial y}[A_h(\frac{\partial v}{\partial x}+\frac{\partial u}{\partial y})]+\frac{1}{H}\frac{\partial}{\partial\sigma}(A_v\frac{1}{H}\frac{\partial v}{\partial\sigma})$$
$$-g\frac{\partial\xi}{\partial y}-\frac{gH}{\rho_0}\frac{\partial}{\partial y}[\int_\sigma^0\Delta\rho d\sigma']+\frac{g}{\rho_0}\frac{\partial H}{\partial y}[\int_\sigma^0\sigma'\frac{\partial\Delta\rho}{\partial\sigma'}d\sigma']+fu \tag{3-2}$$

$$\frac{\partial w}{\partial t}+u\frac{\partial w}{\partial x}+v\frac{\partial w}{\partial y}+\frac{\omega}{H}\frac{\partial w}{\partial\sigma}=-\frac{1}{\rho_0}\frac{1}{H}\frac{\partial p_n}{\partial\sigma}+\frac{\partial}{\partial x}(K_h\frac{\partial w}{\partial x})+\frac{\partial}{\partial y}(K_h\frac{\partial w}{\partial y})+\frac{1}{H}\frac{\partial}{\partial\sigma}(K_v\frac{1}{H}\frac{\partial w}{\partial\sigma}) \tag{3-3}$$

$$\frac{\partial\xi}{\partial t}+\frac{\partial}{\partial x}[\int_{-1}^0 Hud\sigma]+\frac{\partial}{\partial y}[\int_{-1}^0 Hvd\sigma]=0 \tag{3-4}$$

where (u,v,ω) are the velocity components in σ coordinates; t is time; ρ is fluid density and ρ_0 is reference density; p_n is atmospheric pressure at the free surface; H is water depth; ξ is surface elevation; g is gravitational acceleration; f is the Coriolis parameter; A_h, A_v are horizontal and vertical eddy viscosity coefficients; K_h, K_v are horizontal and vertical eddy diffusivity coefficients; horizontal eddy viscosity and diffusivity are obtained by the Smagorinsky sub-grid turbulence model *(Smagorinsky, 1963)* and vertical eddy viscosity and diffusivity are given by the k-ε model *(Mellor and Yamada, 1982; Warner et al., 2005)*; and w is vertical velocity component in the z direction. The dependence between w and ω reads,

$$\omega=w-u(\sigma\frac{\partial H}{\partial x}+\frac{\partial\xi}{\partial x})-v(\sigma\frac{\partial H}{\partial y}+\frac{\partial\xi}{\partial y})-(\sigma\frac{\partial H}{\partial t}+\frac{\partial\xi}{\partial t}) \tag{3-5}$$

Following *Casulli and Walters (2000)* and *Zhang and Baptista (2008)*, a semi-implicit scheme is used to obtain an efficient numerical algorithm whose stability is independent of the gravity wave term. The advection terms in the momentum equations are solved using an Eulerian-Lagrangian back-tracking method *(Oliveira and Baptista, 1995; Hu et al., 2009a)* to avoid Courant number constraint. In addition, in order to release the constraint of orthogonality of the grid, most terms in momentum equations are evaluated at the centre of the cell using a finite volume scheme, and then interpolated into the face in a conservative way *(Fringer et al., 2006)*.

For the purpose of modeling the whole water exchange and tidal propagation of the Yangtze Estuary, which is controlled both by tidal and fluvial influences, the computation domain is as large as 620 km by 680 km, and it contains the entire Hangzhou Bay, part of East China Sea and Yellow Sea and the whole tidal zone of Yangtze River. The horizontal and vertical meshes are shown in *Figure 3-12*. The sea boundary (pink lines in *Figure 3-12*) and river boundary (red dot in *Figure 3-12*) are specified by 16 tidal components and observed river discharges at the Datong. More details about the model descriptions and calibrations are given in *(Qi, 2007)*.

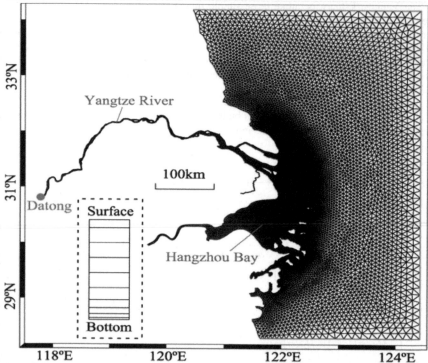

Figure 3-12. Vertical and horizontal computational mesh. Vertical 10-σ layers are divided by 2%, 4%, 6%, 8%, 12%, 15%, 15%, 15%, 15% and 8% thickness of water depth from bottom to surface, which is shown in the dashed box. Horizontal triangle grids are hybrid fitting to local engineering constructions in Figure 3-1 and with fine resolution near the river plume (the red dot and pink lines are the locations of river and sea boundary, respectively).

Residual currents in tidal estuaries and coastal embayments have been recognized as fundamental factors that affect the long-term transport processes *(Liu et al., 2011)*. It has been pointed out by previous studies *(e.g. Zimmerman, 1979; Feng et al., 1986)* that it is more relevant to use Lagrangian residuals than Eulerian residuals to determine the movements of masses. The definition for the Lagrangian residual current is the net displacement of a marked water parcel divided by the lapsed time.

$$\begin{cases} u_L = (\frac{1}{T}\int_0^T (u_i h)dt) / h_0 \\ v_L = (\frac{1}{T}\int_0^T (v_i h)dt) / h_0 \end{cases}$$

(3-6)

where (u_L, v_L) are Lagrangian residual velocity components; (u_i, v_i) are horizontal velocity components of each layer; t is time; T is the selected tidal cycle period (here the value is 24 h and 50 min *(Chu et al., 2010)*); h is time-dependent water depth; and h_0 is tidal-averaged water depth.

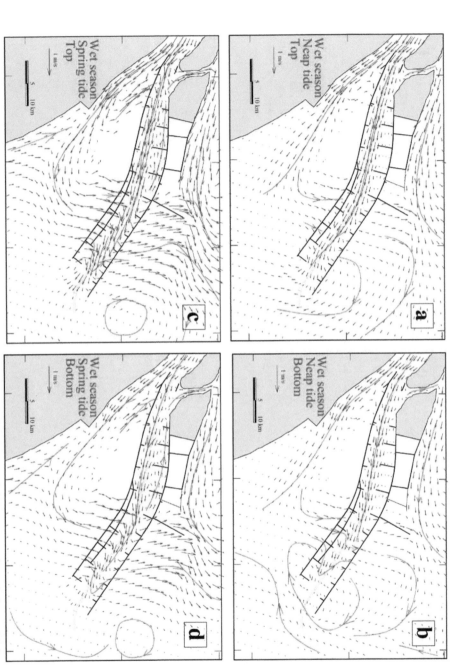

Figure 3-13. Tidal period-averaged surface (a, c, e and g) and bottom (b, d, f and h) Lagrangian residual current fields in respect of neap (a, b, e and f) & spring (c, d, g and h) tide phases and wet (a-d) & dry seasons (e-h). The bold pink lines are the dominant transport directions.

Multiscale physical processes of fine sediment in an estuary

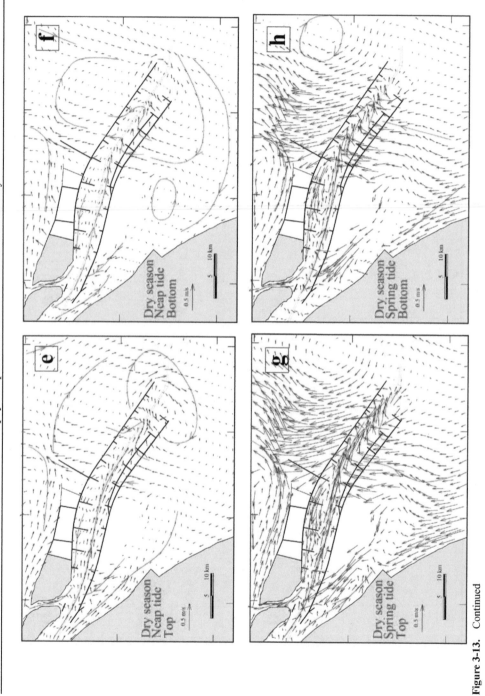

Figure 3-13. Continued

Figure 3-13 shows the tide-induced Lagrangian residual current fields for different conditions in the North Passage. Simulations are carried out in both spring and neap tides in typical wet and dry seasons. From these residual current fields, some fundamental information could be drawn out from the comparisons between surface-bottom layers, spring-neap tides and wet-dry seasons.

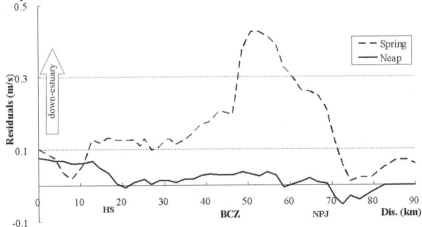

Figure 3-14.   Comparisons of along-channel components of residuals in the bottom layer during neap and spring tides; it indicates that it is much easier for suspended sediments to be flushed away from the channel during spring tides than during neap tides. Note that the positive direction of the residual is down-estuary while the negative direction is up-estuary).

Numerical results (*Figure 3-13*) show that the fluvial excursions (down-estuary direction) in the DNC are dispersed near the lower reach by the tidal currents from JDS. It suggests the water mass can not be transported easily outside the channel. At the head of the DNC, where there are seemingly more submarine tributaries than in other reaches (see *Figure 3-2*), the up-estuary residuals from the South Passage (the South Passage is considered as the main sediment outlet *(Mao et al., 2001)* of the Yangtze Estuary) and the shoals in the channel-north side would disturb and weaken the down-channel transport trend. A comparison between surface and bottom residual fields shows that the down-channel tendency in the top layer is greater than in the bottom layer, which is also not beneficial for the drainage and mixture of near-bed high-concentration suspensions. In addition, the simulated hydrodynamical results (*Figure 3-13*) show that sediments during neap tide would be less active and be more prone to deposition than during spring tides in view of sediment availability.

Details about the along-channel components of the Lagrangian residuals during neap and spring tides are presented in *Figure 3-14*. As we can see clearly, the magnitudes of along-channel residual during neap tides are much smaller than those during spring tides. It means that it is very hard for sediments to move out of this reach during neap tides compared to spring tides. According to the average Lagrangian residuals at the upper reach, it would take more than 6 days for those suspended sediments to be displaced to the lower reaches during neap tides, while the time period will decrease to an estimated 1-2 days during spring tides. It is also suggested that if fluid mud is generated, there is only a weak residual transport transferring these materials down-channel in the upper reach, so the fluid mud would just oscillate with the tidal currents in this reach. According to the modeling, the average near-bottom residual current velocity is less than 0.2 m/s. So, in a single flood or ebb tide, without taking into account damping effects near the bed, fluid mud can only move forward less than 5 km , which is smaller than the length of the upper reach of the channel. In short, the tidal hydrodynamics is not favorable for moving the fluid mud outside of the upper reach. Further, the simulated results also confirm that the fluid mud dynamics processes are not only a tidal energy influenced phenomenon but also an advective phenomenon.

3.4 Discussion

3.4.1 Process-oriented methods

It is still unclear how the fine sediment-current interaction would affect the basic erosion-deposition process near the bed layer. However improved understandings have been made by detailed field measurements and laboratory experiments *(e.g. Ali and Geoprgiadis, 1991; Le Hir et al., 2000; McAnally et al., 2007a; Manning et al., 2010)*. In this study, we examine the physical processes of fluid mud (i.e. fluidization, flocculation and downslope sliding) aiming at explaining the dynamics of storm-induced fluid mud in a muddy-estuarine navigational channel.

3.4.1.1 Fluidization & sediment boundary layer

Wave energy would loosen and break initial bed matrix (this is called fluidization), and release those sediment particles into the upper water column. The fluidization process is easy to understand but it is very hard to quantify in different bed compositions and wave conditions. Two empirical formulas from the literature *(Madsen and Wood, 2002; van Rijn, 2007)* are introduced to estimate the thickness of the sediment mixing layer at the peak of the storm (δ_{cw}) in *Table 3-2*. From the table, we can see that the predicted thickness of possible liquefied bed layer is not able to explain the observed thickness of fluid mud layer.

Table 3-2. Comparisons of different empirical formulas of thickness of sediment mixing layer.

Formula	Reference	Parameters	Results (m)
$\delta_{cw} = \dfrac{\kappa u_{*m}}{\omega}$	*(Madsen and Wood, 2002)*	H_s=0.5-2 m ω=1.4-1.8	0.02-0.08
$\delta_{cw} = 2\gamma_{br}\delta_w$ with limits $0.1 \leq \delta_{cw} \leq 0.5$ m	*(van Rijn, 2007)*	$f_w = f_{cw} = 0.01$ h=2-15 m	<0.12

Note: $\kappa = 0.4$ is the von Karman's constant; u_{*m} is the maximum combined shear velocity; ω is wave radian frequency; $\delta_w = 0.36 A_\delta (A_\delta / K_{s,w,r})^{-0.25}$ is the thickness of wave boundary layer; A_δ is the peak orbital excursion based on significant wave height (H_s); $K_{s,w,r}$ is the wave-related bed roughness; $\gamma_{br} = 1 + (H_s / h - 0.4)^{0.5}$ is an empirical wave breaking coefficient ($\gamma_{br} = 1$ for $H_s / h \leq 0.4$); h is the water depth.

3.4.1.2 Flocculation & hindered settling

Flocculation is considered a significant sedimentary process in muddy estuaries, and it also interacts with and results in other profound processes, such as turbulence damping *(Le Hir et al., 2000)* and hindered settling. *Manning et al. (2010)* and *Shi (2010)* reviewed the relationships between floc size, mean concentration and settling velocity. They show the significant difference in fall velocity between single particles and flocs. In general, the fall velocity of flocs is of the order of 10 times larger than that of an individual particle. From some point-sampled data, the net settling velocity is estimated ranging from 0.4 to 4.1 mm/s *(Shi, 2010)*. Determination of settling velocity is a key to understanding and modeling cohesive sediment transport.

From *Figure 3-14*, the first stage during the dependency of settling velocity and sediment concentration is free settling, and then it is flocculation settling. The settling velocity will be accelerated significantly with increasing of sediment concentration in the second stage. The fall velocity would not increase infinitely as the concentration increases; hindered settling is followed after flocculation process at some critical point, see *Figure 3-15*. Hindered settling is the influence of neighbouring particles on the settling velocity of an individual particle within a suspension *(Winterwerp, 1998)*. This characteristic of fine suspended sediment explains the vertical trapping effect near the bed and the maintenance status of fluid mud.

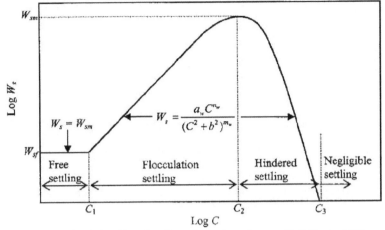

Figure 3-15.   Schematic relationships of settling velocity dependence on SSC *(Costa, 1989)*.

3.4.1.3   Downslope transport of fluid mud

The issue of downslope transport of fluid mud has received considerable attention in recent years. Converging effect *(Fuhrman et al., 2009)* from a sloping bed and trapping effect *(Jensen and Fredsoe, 2001)* on dredged channel can potentially create net cross-trench transportation of sediments. *Ali and Geoprgiadis (1991)* derived a theoretical formula for the downslope velocity of fluid mud from the momentum and continuity equations, and verified it by subsequent laboratory experiments. *Traykovski et al. (2000)* derived a Chezy equation from the balance of gravitational forcing and friction (see *Figure 3-16*), and validated it by observations on the Eel River continental shelf. The two formulae are applied to estimate the downslope velocity of fluid mud at the flank of dredged channel at DNC. As shown in *Table 3-3*, the results indicate that the cross-channel motion of fluid mud at a slope is very considerable at a destined direction, and it would behave independently of the upper oscillating fluid. Therefore, that process would trap the fluid mud from nearby shoals gradually. This may be an explanation for the fact that the fluid mud is almost only observed in the excavated region of the North Passage, as shown in *Figure 3-8*.

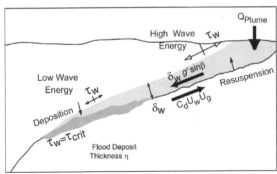

Figure 3-16.   Schematic downslope movement of fluid mud under wave function *(Traykovski et al., 2000)*.

Table 3-3. Comparisons of downslope velocity of fluid mud by different empirical formulae.

Formula	Reference	Parameters in this case	Results (m/s)
$u_d = \left(\dfrac{8\Delta\rho_m g d_m S - 8\tau_B}{\rho_m(1+\alpha) f_w} \right)^{1/2}$	(Ali et al., 1997)	$\alpha = 0.43$; $d_m = 0.1-0.2m$; $g = 9.8$ N/kg; $\rho_m = 1100$kg$/$m^3; $\Delta\rho_m = 80$ kg$/$m^3;	0.1-0.7
$d_m g'S = C_d u_d^{\,2}$	(Traykovski et al., 2000)	$\rho = 1600$ kg$/$m^3; $\tau_B = 0.1-0.2$ N/m^2; $S = 0.001-0.02$; f_w=0.01; C_d=0.003; $g' \approx 0.5$ N/kg	0.1-0.6

Note: u_d is downslope velocity of fluid mud; $\Delta\rho_m$ is average density difference between fluid mud layer and upper suspensions; g is gravitational acceleration; d_m is the thickness of fluid mud; S is slope; τ_B is the Bingham yield stress; ρ_m is the density of fluid mud; α is an empirical coefficient; f_w is a wave-related friction factor; $g' = g\dfrac{\Delta\rho_m}{\rho}$ is reduced gravity; ρ is the bulk density of settled bed; C_d is the drag coefficient.

3.4.2   Engineering-oriented methods

From the above analysis, it seems unlikely that those local micro-processes could explain the sudden increase in fluid mud, we turn to large-scale processes. The following two numerical tests are carried out to answer two basic questions: (1) where does the sediment come from, and (2) is it possible for those sediments to contribute to providing "raw materials" for fluid mud (similar to (McAnally et al., 2007a)).

3.4.2.1   Sediment availability

The presence of wave motion with sufficient dissipation energy in shallow shoals could entrain and load a large number of surficial sediment particles on the riverbed (Jing and Ridd, 1996). If the SSC increases sharply on the spot, then the high turbidity water-sediment mixtures would be carried away by subsequent tidal currents. There is then a strong tendency to support and transport them to those locations with lower wave-current shear stresses. So it can be expected that those tidal flats near the North Passage are possible sediment sources during wave and storm times.

Here, we can estimate stress by our numerical test and determine a direct relationship between bed shear stress and sediment supply, which indicates that the potential for erodible sediments is determined by intensity of bottom shear stress.

The wave-induced shear stress (τ_w), current-induced shear stress (τ_c) and combined shear stress (τ_{cw}) are calculated as following Madsen and Wood (2002) and Dufois et al. (2008) expressions 3-7~3-9. The results at the peak of the storm are presented in Figure 3-17.

$$\tau_c = \rho u_{*c}^{\,2}, \text{ with } u_{*c} = \frac{\kappa u(z)}{\ln(z/z_0)} \tag{3-7}$$

$\tau_w = 0.5\rho f_w u_b^{\,2}$, with $f_w = 0.3$

If $A/k_s < 1.57$;and beyond: $f_w = 0.00251 \exp(5.21(A/k_s)^{-0.19})$ (3-8)

$$\tau_{cw} = [(\tau_m + \tau_w|\cos\varphi|)^2 + (\tau_w \sin\varphi)^2]^{0.5}, \text{ with } \tau_m = \tau_c\left[1 + 1.2(\frac{\tau_w}{\tau_w + \tau_c})^{3.2}\right] \tag{3-9}$$

where ρ is the density of water; $\kappa = 0.4$ is Von Karman's constant; u_{*c} is bottom friction shear velocity; $u(z)$ is near-bottom velocity; z is the height of the first layer above the bottom; z_0 is the bed roughness length; f_w is wave-related friction factor; u_b is wave orbital velocity; A is orbital half-excursion near the bottom ($A = u_b T / 2\pi$) based on u_b and significant wave period (T).

From *Figure 3-17*, it is clearly shown that wave and current have significantly different effects on bottom shear stress. In *Figure 3-17a*, it shows that the stress induced by currents alone at the lower reach of the North Passage is stronger with respect to the upper reach (the critical erosion stress is about 0.1-0.8 N/m^2 *(Hu et al., 2009b)*, and it also means that the fluid mud generated or located at the lower reach would be easy to resuspend. In *Figure 3-17b*, one can see that the wave function cannot act directly along the DNC; it is only witnessed at those shallow shoals, such as the Hengsha Shoal, (HSS), the Jiuduansha Shoal (JDS), the Nanhui East Shoal (NHES) (their locations are shown in *Figure 3-2*), and some dike-shielded flats in the North Passage; that means those areas might provide available sediments to nearby water columns during suitable energetic wave conditions. The combined effects of wave and current on bottom shear stress can enhance the erosion capacity (*Figure 3-17c*). Time-series of shear stress (*Figure 3-18*) of some sample points (locations can be referred from *Figure 3-9*) at the same sites are extracted from the map files to show the relationship between wave variation and bottom shear stress. From *Figure 3-18* it is obvious that without the wave function, the magnitude of bottom shear stress on the shoals is very low; it means that in regular weather, those areas would favor deposition rather than erosion but during strong waves, the deposited sediment may be entrained and loaded by wave dissipation *(Jing and Ridd, 1996)*. In the lower panel of *Figure 3-17*, the horizontal distribution of combined bottom shear stresses is depicted by color contour map. The warm color tone, which means areas of high possibility of erosion, is mainly situated at the lower reach of the DNC and other adjacent shoals; and the stress in the observed fluid mud zones is obviously smaller than other reaches. It further indirectly illustrates the fact that the fluid mud, once existing in the upper reach of the North Passage, would be less erodible in comparison with that at the lower reach of the DNC.

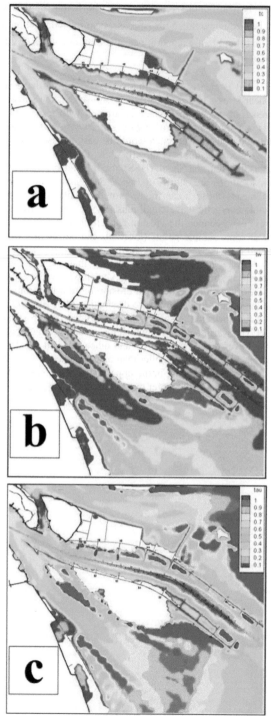

Figure 3-17. Horizontal distribution of modelled bottom shear stress by: (a) current; (b) wave; and (c) combined wave and current at the peak of the storm, unit: Pa.

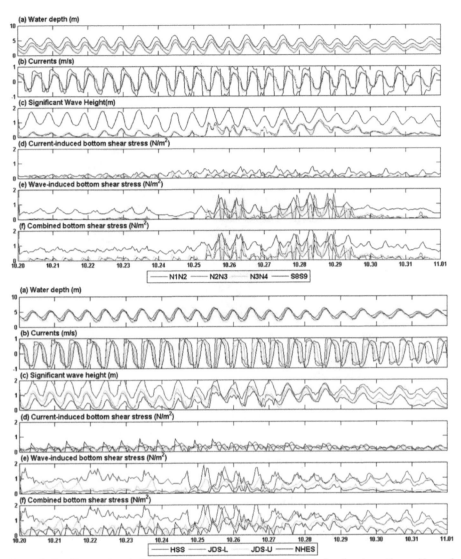

Figure 3-18. Temporal evolution of bottom shear stress at the shoals during the storm (the positions of sample points are shown in *Figure 3-9*).

Moreover, for the 2DH wave model (SWAN), wave breaking is simplified by depth-induced wave breaking mechanism. So a simple coefficient, the parameter γ *(Battjes and Stive, 1985; Ruessink et al., 2003)* is used as the criteria for wave breaking. The value in this study case is constant, $\gamma = H_m / d \approx 0.68$ (H_m is peak wave height, d is water depth). On the other hand, according to the measurement data at the DNC (the location is shown in *Figure 3-2*), the ratio of peak wave height to significant wave height is about 1.6, $H_m / H_{1/3} \approx 1.6$. So at the critical wave breaking conditions, $\alpha = H_{1/3} / d \approx \gamma / (H_m / H_{1/3}) = 0.68 / 1.6 \approx 0.43$. Actually, it is not possible to obtain the original wave heights in those wave breaking areas from the simulated results, because the wave has broken already, and modeled wave heights are reduced judging by wave breaking condition (γ). As we know in coastal and estuarine areas, wave energy mainly dissipates sharply in breaking zones, and a part of this energy is consumed by near-bottom sediment movement (erosion and suspension). Therefore, the sediment in the wave breaking area has high potential for participating in suspended sediment transport, and it is considered as a sediment source by wave action. In order to quantify the area of wave breaking zone, a non-dimensional parameter, the wave breaking factor (η_w), is defined as.

$$\eta_w = H_{1/3} - \alpha d \tag{3-10}$$

in which $\alpha \approx 0.43$ is the wave breaking coefficient. when $\eta_w > 0$, wave breaking occurs.

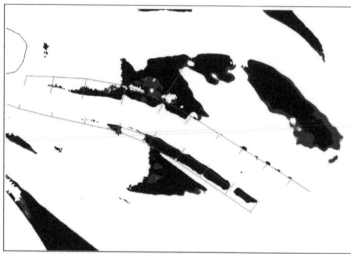

Figure 3-19. Simulated wave breaking area nearby DNC.

In *Figure 3-19*, the possible wave breaking areas are displayed in a dark color determined by the value of η_w at the peak of the storm. These results nearly coincide with the horizontal distribution of modeled bottom shear stress. It also shows wave breaking zones situated nearby the North Passage during this fluid mud event.

3.4.2.2　Sediment transport pathway

From above analysis it is shown that, during this winter storm, extensive tidal flats and shoals could be disturbed by waves; but as we can see, those areas are far from the observed fluid mud locations, so, in this part, the fate or probable destinations of those loaded sediments need to be examined.

To identify if it is possible for those sediments stirred up from ambient shoals to be transported into the nearby the RMB areas, see *Figure 3-7*, the pathways of sediment transport are estimated at several locations (see *Figure 3-9*) near the DNC. At each location, the sediment particles are assumed only to be transported near the bed layer, similar to the motions of near-bed high-concentrated suspensions, which have just been eroded from the fluid mud layer. The driving force is only related to hydrodynamics, but the wave function is also simply taken into account in a simple way, that is, the incipient velocity is decreased to 0.5 m/s (ordinarily, without considering wave agitation, the incipient velocity is always 0.8-1.0 m/s *(Cheng et al., 2003)* in this area). By the way, for comparison, the pathways of the suspended sediment during spring and neap tides are simulated. The results are shown in *Figure 3-20*.

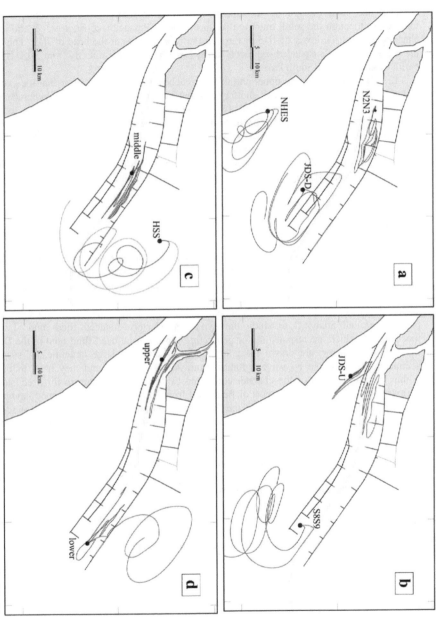

Figure 3-20. Simulated sediment transport pathways (the red lines and blue lines mean pathways during flood and neap tides, respectively). (a) Initial release locations are at the points N2N3, JDS-D and NHES. (b) Initial release locations are S8S9 and JDS-U. (c) Initial release locations are HSS and at the middle reach of the DNC. (d) Initial release locations are at the upper and lower reaches of the DNC.

In *Figure 3-20*, three obvious characteristics of the pathways of eroded suspended sediments can be found as follows.

(1) Due to the effect of the Coriolis force in this region, the direction of tidal rotating flow is clockwise outside the North Passage, so the overall transport direction of mass is from north to south. JDS is the first transitional stop of suspended sediments from HSS and the North Passage; and with the aid of subsequent flood tidal currents, those sediments would be transported theough the south training dike at the North Passage (see *Figure 3-2*) and into the channel either at the lower of JDS or through the watercourses at the head of JDS. This means the main sources of suspended sediments at the North Passage are directly from JDS, HSS, the North Passage and the South Passage.

(2) The pathways in the upper and middle reaches of the channel show that the excursion distance of tidal currents is much shorter than that in the lower reach, which generally agrees with the results of Lagrangian residuals (*Figures 3-13* and *3-14*). It indicates that the sediments transported from the head of JDS or the South Passage into the North Passage will be prone to linger at the upper and middle reaches, while in the lower reach during flood tide, they could easily be transported to far away from the DNC.

(3) Sediments at HSS, JDS (both the head and the lower parts), shallow areas shielded by groins and the North Passage are the main sources during this fluid mud event.

3.5   Conclusion

A storm-induced fluid mud event is investigated in a muddy-estuarine navigational channel from the views of both process-oriented and engineering-oriented approaches by the means of observations, hydrodynamic modeling and wave propagation simulation. A non-dimensional parameter, wave breaking factor, is proposed in this chapter to investigate the wave-induced sediment availability. Besides, the Lagrangian residuals, bed shear stress and sediment transport pathway are all analyzed to study the dynamics of storm-induced fluid mud. Two key mechanisms, which are responsible for generating the storm-induced fluid mud in the DNC of the Yangtze Estuary, are drawn from this study directly: (1) large quantities of suspended sediments are available by wave agitation from adjacent shoals, and they have to oscillate within the upper reach of the channel controlled by local hydrodynamics for a considerable period; and (2) downslope transport of fluid mud from the flanks to the deepened navigational channel possibly reduces diffusion of the fluid mud, and provides large quantities of sediment sources for fluid mud; this effect may be enhanced with the channel deepening. Modeled and observed results demonstrate that (1) the fluid mud dynamics process is an advective phenomenon determining the central position of fluid mud layer along the channel, and also it is a tidal energy-influenced phenomenon, controlling the dissipation or growth of fluid mud; and (2) both suspended particulate matter availability and local residual flow regime are of critical importance for the trapping probability of sediment and the occurrence of fluid mud.

In addition, other sedimentary processes, such as flocculation and hindered settling, are likely beneficial for the formation and maintenance of fluid mud, but it needs more efforts to improve and transfer the poor theoretical understanding of complex interactions between currents, sediment and other forces near the bed in the channel.

ETM dynamics and saltwater intrusion

<u>Highlights</u>

(1) We explain the development of an ETM in the Yangtze Estuary by analysis of unique data.

(2) Salinity-induced baroclinic pressure gradient forces are a major factor controlling the variation of the vertical velocity structure.

(3) Internal tidal asymmetry is important in maintaining a stable density stratification interface near the estuarine front.

4.1 Introduction

It is generally accepted by coastal scientists and engineers *(Wolanski et al., 1988; Uncles et al., 1990; Jay and Musiak, 1994; Winterwerp, 2006; Manning et al., 2010; Mehta, 2014)* that the occurrence of flocculation and stratification in fresh-salt water mixing regions (salt fronts) makes the dynamics of fine sediment in a muddy estuary a complex process. Understanding the competition between vertical mixing and stratification is of critical importance for characterizing suspended sediment behavior in an estuarine turbidity maximum (ETM) area. At the same time, knowledge of non-linear interactions among tidal force, wind waves, freshwater inflow, the Coriolis force and local bathymetry also contributes to determination of the spatial and temporal variability of saltwater intrusion, sediment availability and ETM dynamics. Vertical micro-scale saltwater-mud interaction coupled with longitudinal macro-scale gravitational circulation produces specific hydrodynamic and mud dynamics. Previous works showed that micro-scale and macro-scale physical mechanisms affect ETM formation including flocculation *(Dyer and Manning, 1999; Xu et al., 2010)*, tidal resuspension *(Shi, 2010)*, settling/scour lag *(Li and Gust, 2000; Pritchard, 2005)*, stratification *(Winterwerp, 2006)*, turbulence damping (suppression) *(Geyer, 1993)*, drag reduction *(Li and Gust, 2000)*, baroclinicity effects *(de Nijs, 2012)*, internal tidal asymmetry (asymmetric tidal mixing) *(Jay and Smith, 1990; Jay and Musiak, 1994; Scully and Friedrichs, 2007; Cheng et al., 2011)*, and external tidal asymmetry (otherwise known as tidal pumping, tidal straining or gravitational circulation) *(Wolanski, 1995; Li and Zhong, 2009; Liu et al., 2011; Winterwerp, 2011a; Jiang et al., 2013a)*. In nature, one particular mechanism may influence another, such that each of these factors may not be fully independent. Furthermore, the dominant mechanism responsible for forming an ETM may vary at different reaches within the same estuary, depending on the condition of each location.

The Yangtze Estuary, located in the eastern China, has previously been classified as a partially mixed estuary *(Chen et al., 1999; Shi, 2010; Guo and He, 2011; Zhang et al., 2012a)*. However, owing to the navigational channel development in the Yangtze Delta area *(Dai et al., 2013; Jiang et al., 2013a; Wan et al., 2014b)*, the North Passage Deepwater Navigational Channel (DNC) has experienced a ten-year channel narrowing and deepening process. Field measurements investigating the flood-ebb and neap-spring variations of saltwater intrusion and ETM dynamics in the DNC were carried out in August 2012. Amongst other phenomena, this study identified an enriched stably stratified flow in this region for the first time. It is known that strong stable stratification is easier formed in a relatively deep environment such as a deep sea *(Komori and Nagata, 1996)*. Only under laboratory enforced conditions *(Stillinger et al., 1983; Itsweire et al., 1986)*, a stably stratified flow in water is observed. So this case is a real-world example for us to understand the physics of the stably stratified flow in shallow water.

In this chapter, we report the measured spatial and temporal data describing the ETM front and salt wedge dynamics in the navigational channel of the Yangtze Estuary, and subsequently investigate the underlying physical mechanisms responsible for the formation of the stably stratified density interface.

4.2 Methods

4.2.1 Field observations

Ten smart hydrological vessels were employed during the field survey, with each individual boat assigned to collect data at one of ten different stations located along the channel (NGN4, CS0, CS9, CS2, CS6, CSW, CS3, CS7, CS4 and CS10 from upstream to downstream; *Figure 4-1*). These vessels mainly collect vertical distribution data of horizontal velocity, SSC and salinity under neap tide (12 August 2012) and spring tide (17 August 2012) conditions during the 2012 wet season (high freshwater inflow). These data are collected hourly at six layers in water column with relative depths of 0.05(near surface), 0.2, 0.4, 0.6, 0.8, and 0.95 (near-bed) at each station. The currents data are obtained by rotor current meter, and SSC and salinity are derived by water sampling method. Strict local maritime administration and the frequent thoroughfare of shipping traffic in the Yangtze Estuary do not allow anchorage in the middle of the navigational channel for a long observational period. For such reason, all previous studies performed in this region have used measurement sites located outside of the channel *(Liu et al., 2011; Dai et al., 2013; Jiang et al., 2013a)*. This work presents a set of hydrological data that have been obtained within the middle of the DNC.

The field campaign is completed using a strict methodology. Each surveying boat stays anchored in a waiting position ~200 meter away from a particular station. Once visibility and safety permissions are granted, the boat moves out into the middle of the channel, and immediately begins data collection. After approximately 10 min, the boat is required to return to the flank of the DNC and resumes its waiting position. All measurements are performed during the daytime for safety reasons. The detailed measurement time in neap and spring tides are presented in *Figure 4-2*. The observation time is from 08:00 to 21:00, 12-8-2012, and spring tide time is from 05:00 to 18:00, 17-8-2012.

It should be noted that the duration of an entire tidal cycle in the Yangtze Estuary is about 25 h *(Qi, 2007)*, which includes two ebb-flood tidal cycles. However, given that the fieldwork was only possible during daytime, the available measured data only span 13-h-long, which represents half of a tidal cycle, see *Figure 4-2*. These data therefore have the limitation that they introduce a small deviation (~0.5 cm/s) for residual currents. The deviation is evaluated from the data near the station CSW (see *Figure 4-3*). A mean freshwater inflow of ~56,000 m³/s during the observation (12-18 August 2012) was recorded at the Datong station (data available online: *http://yu-zhu.vicp.net*), which is regarded as the tidal limit of the Yangtze Estuary *(Guo and He, 2011; Ma et al., 2011)*, and is located about 600-700 km away from the mouth of the estuary.

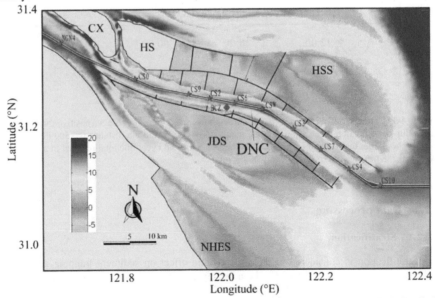

Figure 4-1. Bathymetry with constructed dikes and groins at the North Passage of 2012, all the topography, elevation and water depth in this study are referred to the Theoretical Depth Datum. CX and HS are two fixed boundary islands Changxing Island and Hengsha Island; JDS, HSS and NHES denote Jiuduansha Shoal, Hengsha Shoal and Nanhui East Shoal respectively; the filled pink diamond (BCZ) is the tide gauge for water level; the filled red stars are the ship-born stations for currents, salinity and SSC, the thin white lines show the 12.5-meter Deepwater Navigational Channel (DNC).

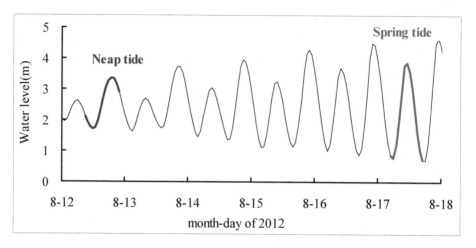

Figure 4-2. Detailed measurement time period coupling with the measured time-series water level at the BCZ station, where is shown in *Figure 4-1*. The bold blue line, from 08:00 12-8-2012 to 21:00 12-8-2012, shows the observation times for neap tide; and the bold red line, from 05:00 17-8-2012 to 18:00 17-8-2012, is period of spring tide.

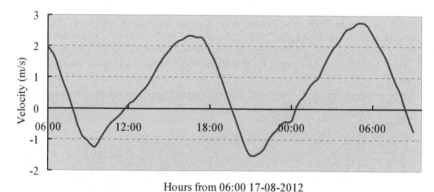

Hours from 06:00 17-08-2012

Figure 4-3. Time series (hourly) of measured depth-averaged current velocity (positive means ebbing) at the ship-borne station (the station is at the south side of CSW and the distance is ~500 m, see *Figure 4-1*). In the figure, the residual is 0.584 m/s in the first half tidal cycle, the total residual in a whole tidal cycle is 0.588 m/s, the difference is less than 0.5 cm/s.

4.2.2 Data processing

4.2.2.1 Stratification

Stratification is considered to have a profound effect on sediment trapping and vertical sediment mixing, and it is thought to be a principal mechanism for the formation of an ETM *(Winterwerp, 2006; Li and Zhong, 2009)*. A widely accepted definition of the estuary stratification parameter that is based on the vertical structure of salinity, was proposed by *Hansen and Rattray (1966)*. They defined the salinity stratification parameter of any specified location, a non-dimensional index, as follows:

$$N = \frac{S_b - S_s}{S_p} \tag{4-1}$$

where N is the salinity stratification parameter, S_b and S_s are the salinity in bottom and surface layer respectively, and S_p is the vertical mean salinity.

The magnitude of the salinity stratification parameter is usually positively correlated with the degree of stratification. For example, a water column with $N>1$ is highly stratified in the presence of a salt wedge, a water column with $0.1<N<1$ is partially mixed, a water column with $N<0.1$ is well-mixed *(Haralambidou et al., 2010)*. It should be noted that a given estuary cannot always be fully characterized by a single parameter. For example, it can be well mixed during the dry season and partially mixed during the wet season or be well mixed in the upper reach and partially mixed in the lower reach *(Nguyen, 2008)*.

Alongside N, the bulk/layer Richardson number (Ri_L), *equation 4-2*, and gradient Richardson number (Ri_g), *equation 4-3* can also be used to determine the state of mixing/stratification of a fluid. These parameters are defined as follows:

$$Ri_L = \frac{gh\Delta\rho}{U^2\rho} \tag{4-2}$$

$$Ri_g = \frac{g\dfrac{\partial\rho}{\partial z}}{\rho(\dfrac{\partial u}{\partial z})^2} \tag{4-3}$$

where g is the gravitational acceleration, h is water depth, U is the depth-mean velocity, $\rho = \rho_0 + 0.78*S + 0.62*C$ is the density of saltwater containing suspended sediments *(Guan et al., 2005)*, $\rho_0 = 1000$ kg/m³ is the reference water density, S is the salinity in practical salinity units (psu), C is the SSC in kg/m³, $\Delta\rho$ is the surface to bottom density difference, $\dfrac{\partial\rho}{\partial z}$ is the vertical density gradient, $\dfrac{\partial u}{\partial z}$ is the vertical gradient of horizontal velocity.

Dyer and New (1986) suggested that Ri_L was less sensitive to the vertical measurement interval and velocity precision than Ri_g. Using measured salinity data, *Dyer and New (1986)* defined three stratification conditions: a stable density stratification interface with no significant mixing $(Ri_L > 20)$, a density interface that is modified by mixing $(2 < Ri_L < 20)$, and strong mixing with no significant stratification $(Ri_L < 2)$. Conversely, fluid tends to be well mixed when Ri_g is less than a critical threshold value (the Kelvin-Helmholtz instability), which has been suggested by many authors *(Abarbanel et al., 1984; Miles, 1986; Geyer and Smith, 1987; Gerz and Schumann, 1996; Peters, 1997; Cudaback and Jay, 2000)* to be in the range ~0.25-1. When $Ri_g > $ ~0.25-1, the vertical velocity gradient is unable to overcome the vertical density difference and the water body is stably stratified.

4.2.2.2 Residuals & tidal pumping

Residual currents in tidal estuaries and coastal embayments were previously recognized as fundamental factors that affect process related to long-term transport *(Liu et al., 2011)*. Previous studies *(e.g. Zimmerman, 1979; Feng et al., 1986)* suggested that the movement of mass is more suitably determined by using a mass transport velocity rather than an Eulerian mean velocity. The definition for the mass transport velocity is the net displacement of a specific water parcel divided by the lapsed time:

$$\begin{cases} u_L = (\dfrac{1}{T}\int_0^T (u_i h)dt)/h_0 \\ v_L = (\dfrac{1}{T}\int_0^T (v_i h)dt)/h_0 \end{cases} \tag{4-4}$$

where (u_L, v_L) are the mass transport velocity components; (u_i, v_i) are the horizontal velocity components of each layer; t is the time; T is the selected tidal cycle period; h is the time-relevant water depth; and h_0 is the tidal-averaged water depth.

4.2.2.3 Estimation of the turbulent eddy viscosity and diffusion

In an estuarine system, the studies *(Jones, 1973; Pacanowski and Philander, 1981; Nunes Vaz and Simpson, 1994)* suggest that the vertical eddy viscosity (A_v) and vertical eddy diffusivity coefficient (K_v) can be calculated as follows:

$$\begin{cases} A_v = \dfrac{A_0}{(1+5Ri_g)^2} + A_b \\ K_v = \dfrac{A_v}{1+5Ri_g} + k_b \end{cases} \tag{4-5}$$

where $A_0 = 5 \times 10^{-3}$, $A_b = 10^{-4}$, $k_b = 10^{-5}$

4.2.2.4 Comparison of barotropicity and baroclinicity effects

The effects of barotropic and baroclinic pressure gradients on water motion are implicit in the momentum balance equation *(de Nijs, 2012)*, so their direct contribution to the structure of currents is not straightforward to understand. In order to determine the dominant term, accelerations induced by the barotropic and baroclinic pressure gradients must be quantified separately for estimating their relative influence on fluid motion. In the horizontal momentum equation, the pressure gradient term is a function of the barotropic and baroclinic pressure gradient terms, as follows:

$$\frac{1}{\rho_0}\frac{\partial p}{\partial x} = \underbrace{g\frac{\partial \xi}{\partial x}}_{E} + \underbrace{\frac{g}{\rho_0}[\int_z^\xi \frac{\partial \rho}{\partial x}dz']}_{F} \tag{4-6}$$

which x is the along-channel direction, ξ is water surface elevation and p is fluid pressure, which includes hydrostatic (barotropic) pressure (E) and the density gradient induced (baroclinic) pressure (F).

4.3 Results

4.3.1 ETM and salt wedge excursions

The along-channel vertical distributions of velocity, SSC and salinity during the spring and neap in the wet season of 2012 are present in *Figure 4-4* and *4-5* respectively. The measurement data show the near-bed SSC reaching a maximum value of ~80 kg/m³, and oscillating between stations CS6 and CS7 (over a distance of ~30 km) during various tidal types and cycles. The data in *Figure 4-3* suggest that the core of the ETM is relatively stable and always oscillates over a distance of about 30 km between stations CS9 and CS7 (positions see *Figure 4-1*), coinciding with the steepest vertical salinity gradient. The 1-psu isoline appears to mark the upper boundary of the ETM core. The salt front cannot be totally displaced seaward outside the channel during a high fluvial discharge. In *Figures. 4-4g, j* and *4-5g, j*, during flooding processes, the magnitude of flood currents near the bottom is greater than that at the surface layer, especially during neap tide. It implies that the tide in the middle reach of the North Passage first receives the flooding energy at the bottom layer, and it seems it is not easy for the flood tide to propagate to surface layer. And the flood tide is even imprisoned by the ebb tide during neap tide. With regard to internal flow structure, the vertical profile of the current is largely modified in the middle reach. Then, it affects vertical and longitudinal sediment and salt transport. Internal tidal asymmetry has been related to the longitudinal straining of the ETM density current by the tidal and residual flows *(Jay and Smith, 1990; Simpson et al., 1990; Peters, 1997)*, which will consequently promote near-bottom salt intrusion.

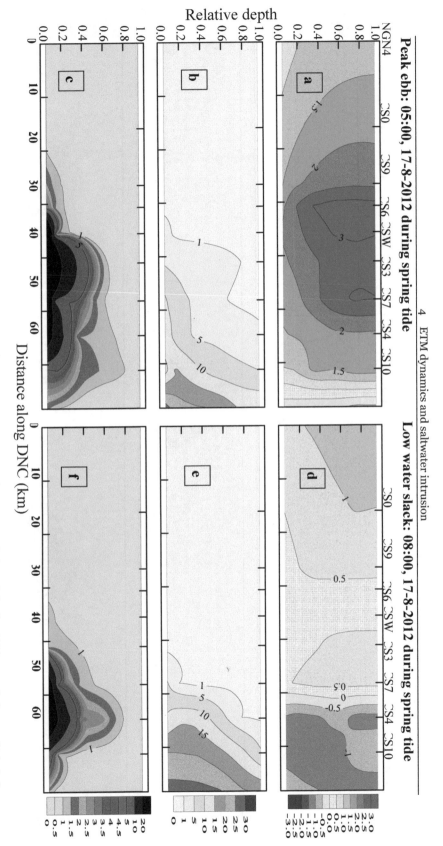

Figure 4-4. Along-channel distributions of velocity, salinity and SSC profiles at the time of peak ebb, low water slack, peak flood and high water slack respectively during spring tide. (a d g h) are velocity profiles (positive ebb) in m/s, red color means a positive velocity, and the color of blue denotes a negative velocity, (b e h k) are salinity in psu, (c f i l) are SSC in kg/m³. (a b c) show the distributions at the peak ebb, (d e f) are at the low water slack. (g h i) show the distributions at peak flood, (j k l) display the distributions at the slack water of spring tide.

Multiscale physical processes of fine sediment in an estuary

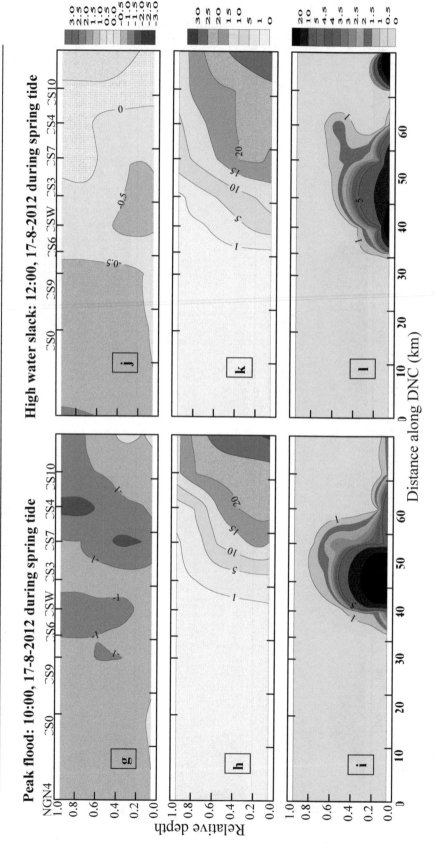

Figure 4-4. Continued.

68

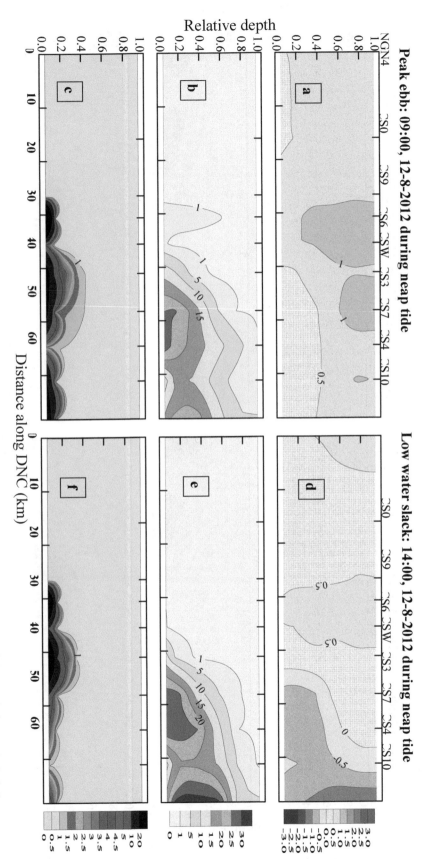

Figure 4-5. Along-channel distributions of velocity, salinity and SSC profiles at the time of peak ebb, low water slack, peak flood and high water slack respectively during neap tide. (a d g h) are velocity profiles (positive ebb) in m/s, red color means a positive velocity, and the color of blue denotes a negative velocity, (b e h k) are salinity in psu, (c f i l) are SSC in kg/m³. (a b c) show the distributions at the peak ebb, (d e f) are at the low water slack. (g h i) show the distributions at peak flood, (j k l) display the distributions at the slack water of neap tide.

Multiscale physical processes of fine sediment in an estuary

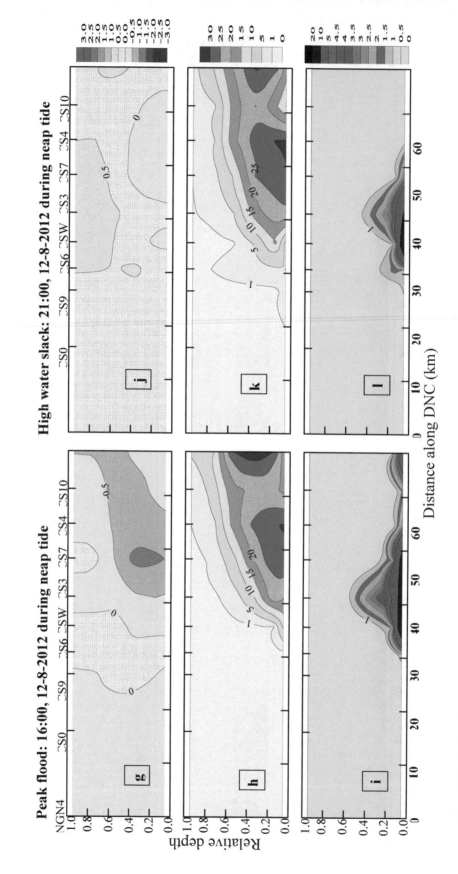

Figure 4-5. Continued.

70

Data show that a salt wedge was formed during both neap and spring tides but it seemed to be much more profound during neap tide than during spring tide (*Figure 4-5*). It did not show significant variation with tidal currents and its position fluctuated between stations CS3 and CS7 (over a distance of less than 10 km). The core of the ETM appeared to remain stationary near station CSW (see *Figures 4-1* and *4-5*), which was located about 10 km further up the estuary from the salt wedge. Furthermore, the internal tides were significantly more asymmetric during neap tide, with the current directions even reversing at the bottom and surface layers during flooding near to the ETM area.

These data suggest that the ETM is quite stable in the channel. The characteristics of the flood-ebb and spring-neap tidal variations of the internal velocity structure, salt wedge and ETM in the DNC allow us to explore the underlying mechanisms responsible for the formation, dynamics and transport. In addition, almost ~70-80% of the annual siltation in the channel occurs near the middle reach *(Liu et al., 2011)*, which is coincident with the excursion zone of the ETM. Thus this study also considers the relevant physical processes involved with sediment trapping.

4.3.2 Stratification variability

To examine the stratification state of the estuarine water under different conditions, three related parameters, N, Ri_L and near-bed Ri_g were determined. *Figure 4-6* shows along-channel variations of these three parameters at peak ebb, Low water slack, peak flood and high water slack during spring and neap tides. According to the stratification criteria described in *Section 4.2.2*, three conclusions can be drawn. Firstly, N was greater than 1 between stations CS2 and CS4 (*Figure 4-1*) during spring tide (*Figure 4-6a*) and between stations CS9 to CS4 (*Figure 4-6b*) during neap tide, suggesting that the water column in these reaches was highly stratified. Secondly, Ri_L had values larger than 20 between stations CS6 and CS4 (*Figure 4-1*) at high water slack of spring tide (*Figure 4-6c*) and at Low water slack, peak flood and high water slack during the neap (*Figure 4-6d*), which indicates a stable density stratification interface with no significant mixing. Thirdly, the near-bed Ri_g values showed an extremely large range from approximately zero to several hundred during both the spring (*Figure 4-6e*) and neap (*Figure 4-6f*) tidal conditions. Additionally, the values in the middle and lower reaches (between stations CS2 and CS10) of the DNC are significantly larger than the threshold value (~0.25-1), which indicates the presence of stable density stratification. Thus, these parameters all suggest that either a highly stratified layer or a conditionally stable density stratification interface formed in the channel.

In order to show that these data are reliable, the validity of these three parameters has been evaluated. As shown in *Figure 4-6*, there is little difference between the values of N for spring and neap tides, and there is small temporal variation. However, *Figures 4-4* and *4-5* show the significant temporal variation of the vertical salinity gradients. The salinity stratification parameter uses a vertical salinity profile to distinguish the general stratification state of a water column, but it cannot show the contribution from suspended sediment, nor can it identify a stably stratified condition. As discussed by *Dyer and New (1986)*, the gradient Richardson number is sensitive to the vertical measurement interval and velocity precision, and so a small deviation of currents could produce a wide range of Ri_g values. Thus, compared to N and Ri_g, the Ri_L is more practical for measurement analysis.

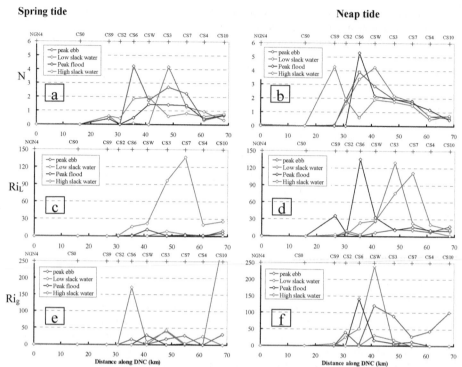

Figure 4-6. Along-channel variations of three stratification parameters at different times (peak ebb, Low water slack, peak flood and high water slack) of the spring (left panels) and neap (right panels) tides. Varied colored lines mean at different times. The upper panels are the variation of salinity stratification parameter, the middle panels are the layer Richardson number, and the lower panels are the gradient Richardson number of the near bed layer. The cross denotes the location of measurement station.

Figure 4-7 shows the temporal variations of Ri_L at station CS3 during the neap and spring tides. These data imply that the stratification effect is conditional, thus it is determined by different tidal types and cycles, which corroborates the conclusion of *Valle-Levinson (2010)*. A stable stratification only appears at the high water slack during spring tide (red line in *Figure 4-7*), whereas stable density stratification lasts for a period of ~6 h between the low and high water slacks during neap tide (blue line in *Figure 4-7*).

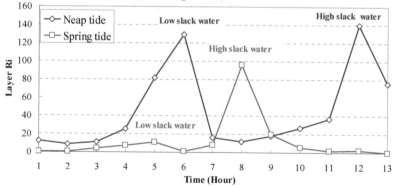

Figure 4-7. Time series of the layer Richardson number at station CS3, the position refers in *Figure 4-1*. The blue line is the variation in neap tide, and the red line is in spring tide.

4.3.3   Current, salinity and SSC residuals

Along-channel distributions of measured bottom and top layers residual currents during spring and neap tides are shown in *Figure 4-8*. This shows that the directions of the near-bottom and near-surface layer residual flows significantly reversed between neap tides at stations CS3, CS7 and CS4. Residuals converged near to stations CSW and CS3 during the neap tide. On the other hand, the near-top and near-bottom residual currents accelerated between stations NGN4 and CS3, and then quickly decelerated between stations CS3 and CS10 (*Figure 4-8*). The longitudinal gradient of residuals changes sharply between stations CS6 and CS7, where the main excursion zone of the ETM occurs.

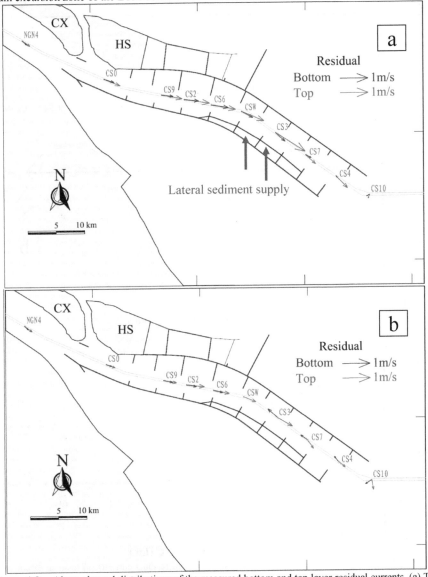

Figure 4-8.   Along-channel distributions of the measured bottom and top layer residual currents. (a) The residuals of spring tide and (b) residuals of neap tide. The red and blue arrows are the residuals at the near water surface and river bed respectively.

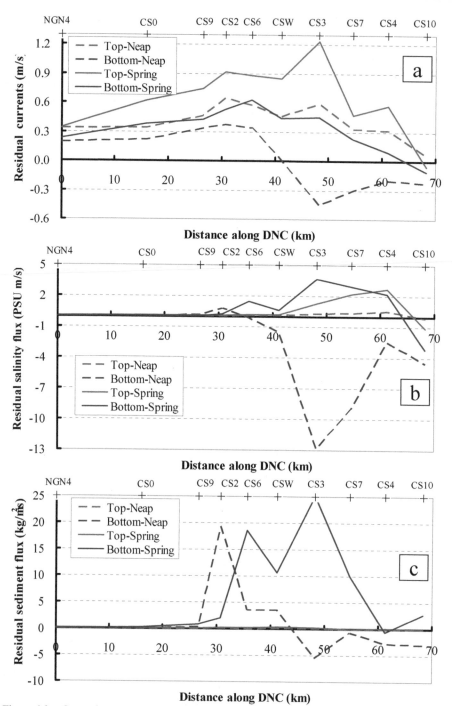

Figure 4-9. Comparisons of the longitudinal distributions of residual currents (a) residual salinity flux (b) and residual sediment flux (c) at the top and bottom layers during neap and spring tides. The dashed lines are residuals during neap tide and the solid lines are residuals during spring tide; the blue lines denote residuals near the bottom and the red lines denote residuals near the top layer. The crosses denote the location of the measurement stations.

Figure 4-9 compares the longitudinal profiles of residual currents, residual salinity flux and residual sediment flux at the top and bottom layers during neap and spring tides, respectively. *Figure 4-9a* shows that the difference between the top and bottom residual currents was largest at station CS3. From the entrance (CS0, ebb-dominant) to the outlet (CS10, flood-dominant) of the North Passage, the intensities of residual flow experience a slowly increasing and fast decreasing process. If this termination (where the magnitude of residuals is ~0 m/s) can move in a seaward direction, the channel would be able to export sediment. Finally, during neap tide, the near-bottom residual currents changed from a seaward transport direction to a landward transport direction just in the middle reach of DNC. This favors the transport of sediments that accumulated downsream to move up-estuary and back into the channel.

Equation 4-4 shows that summation of the product of the relevant salinity or SSC at each time interval produces a residual salinity (*Figure 4-9b*) or sediment flux (*Figure 4-9c*), respectively. In general, the residual sediment flux regime is quite different from the residual currents regime, with the inconsistencies between them attributed to settling lag and scour lag *(Postma, 1961; Pritchard, 2005)*. The overall correlation between residual salinity (*Figure 4-9b*) and residual currents (*Figure 4-9a*) identified in this study agrees with *Jay and Musiak (1996)* and *de Nijs (2012)*, who suggested that the vertical current profile is mainly controlled by salt intrusion. It should be noted that the residual salinity flux at the channel-bottom during neap tide is rather high at the low-reach region of the DNC, indicating that a substantial salt mass accumulation occurred near to station CS3, which is the location of the centre of the salt wedge shown in *Figure 4-5*.

As shown in *Figure 4-9c*, the longitudinal profile of the residual transport is inconsistent with those of both residual current and salinity transport, suggesting the discontinuous sediment delivery along the channel. In addition, there is an apparent increase in the bottom residual sediment flux from stations CS6 to CS3 during the spring tide, and at station CS2 during the neap tide. This indicates that the sediments from other sources are transported into the region (i.e., other than along-channel advective transport). Such possible sources include tidal re-suspension *(Shi, 2010)* and lateral sediment supply, which can be provided by the dike-overtopping currents from the South Passage (depicted by purple arrows in *Figure 4-8a*) during the flooding process. However, the location of the highest residual sediment flux does not coincide with the weakest residual currents, which implies that residual circulation is not the only mechanism for sediment trapping near the ETM area of the Yangtze Estuary.

4.3.4   Baroclinic effect

In a stably stratified estuary, horizontal density differences *(Qiu et al., 1988)* tend to produce a buoyancy force *(Gebhart et al., 1988)*, which is known as baroclinicity (or the baroclinic effect). According to the *equation 4-6*, longitudinal variations of the ratio of the baroclinic (F) and barotropic (E) pressure gradients and the ratio of salinity-induced baroclinic pressure gradient (F_s) and sediment-induced baroclinic pressure gradient (F_c) are shown in *Figure 4-10* for peak ebb, Low water slack, peak flood and high water slack during the spring (left panels) and neap (right panels) tides. The difference between salinity- and sediment-induced baroclinic pressure gradient is that the former calculates the fluid density by the equation $\rho = \rho_0 + 0.78 * S$ ($\rho_0 = 1000$ kg/m^3 is the reference water density, S is salinity in psu), and the latter considers the density is only influenced via SSC ($\rho = \rho_0 + 0.62 * C$, C is the SSC in kg/m^3). And to calculate the total baroclinic pressure gradient, $\rho = \rho_0 + 0.78 * S + 0.62 * C$. It should be noted that F= Fs + Fc.

Multiscale physical processes of fine sediment in an estuary

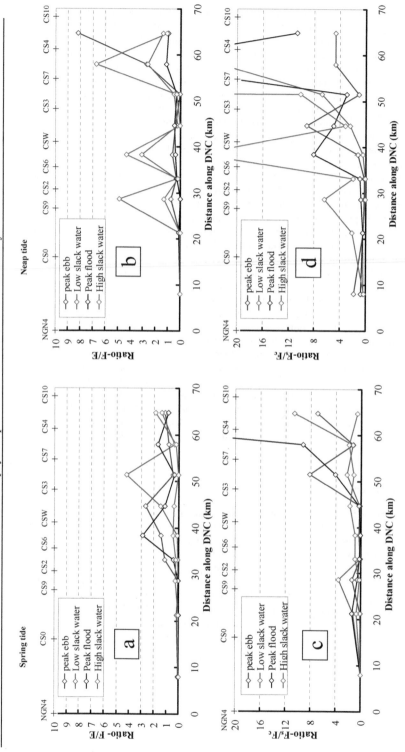

Figure 4-10. Longitudinal variations of the ratio of baroclinic pressure gradient (F) and barotropic pressure gradient (E) in upper panels, and the ratio of salinity-induced baroclinic pressure gradient (Fs) to sediment-induced baroclinic pressure gradient (Fc) (lower panels) in different times (peak ebb, Low water slack, peak flood and high water slack) of the spring (left panels) and neap (right panels) tides. F= Fs + Fc. The cross denotes the location of measurement station.

Figures 4-10a and *b* show that the baroclinic effect had a greater influence in the ETM zone than the barotropic effect, especially during neap tide by the presence of a salt wedge. The salinity-induced and sediment-induced components of the baroclinic pressure gradient were evaluated separately in order to identify which of them was the dominating factor, see *Figures 4-10c* and *d*. The intra-tidal F_s/F_c variations for spring and neap tides showed that the baroclinic pressure gradient was mainly caused by a horizontal salinity-induced density difference, which emphasises the importance of salinity-induced baroclinicity in directly influencing changes in the vertical current velocity structure.

4.3.5   Turbulence damping

The influence of turbulence on the vertical SSC and current profiles has been attracting a great deal of attention *(Jones, 1973; Yamazaki and Osborn, 1990; Nunes Vaz and Simpson, 1994; Peters, 1997; Stacey et al., 1999; Winterwerp, 2006; Scully and Friedrichs, 2007; Cheng et al., 2009; Li and Zhong, 2009; Shivaprasad et al., 2013)*, especially for understanding the fine sediment dynamics *(Geyer, 1993; Jay and Musiak, 1996; Winterwerp, 2006; de Nijs, 2012; Mehta, 2014)*. Turbulence damping/suppression provides great challenges for model simulations aiming to understand near-bed behavior of fine sediment, although a qualitative knowledge is realized that the effect of turbulence damping critically influences the physics of sediment trapping and drag reduction *(Best and Leeder, 1993; Li and Gust, 2000)*. The dependency of the estimated vertical eddy viscosity (A_v) on gradient Richardson number can be calculated from *equation 4-5 (Pacanowski and Philander, 1981)*, and it graphically shown in *Figure 4-11*. It reveals A_v is strongly dependent on Ri_g, especially near the latter's critical threshold value. Thus the vertical eddy viscosity will decrease by one or two orders of magnitude, and the turbulence structure is apparently affected vertically in a stably stratified estuary. By the way, it is needed to note that the empirical coefficient, background (minimum) vertical eddy viscosity (A_b), becomes a sensitive parameter once stratification occurs.

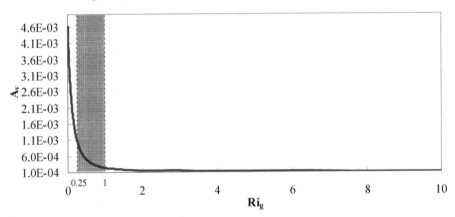

Figure 4-11.   The dependency of vertical eddy viscosity (A_v) on gradient Richardson number (Ri_g), the shaded area is the range of the threshold values of Ri_g.

4.4   Discussion

4.4.1   External tidal asymmetry

The macroscopically hydrodynamic factors in an estuary, related riverine discharge (freshwater inflow), tidal energy, wind climate, the Coriolis force, longshore currents, bed resistance, etc. *(Cudaback and Jay, 2001; Cheng and Wilson, 2008)*, jointly force and generate the phenomenon of external tidal asymmetry. The contribution of each determinant is very complicated and time- and site-specific. In the DNC of the Yangtze Estuary, the variation of tidal energy (the spring-neap response in the wet season) does not alter the flow regime so much that the convergent area (the competition between freshwater inflow and saltwater tidal currents) can be able to displace outside of the channel. So this observation can not significantly show the importance of external tidal asymmetry in maintaining the salt and sediment balances.

4.4.2   Internal tidal asymmetry

Internal tidal asymmetry induced by stratification and baroclinicity only influences local vertical current structure *(Jay and Musiak, 1996)*. *Figure 4-12* shows plots of the measured velocity against the corresponding water depth at the upper (CS0), middle (CS3) and lower (CS10) reaches of the DNC during the spring (left panels) and neap (right panels) tides. The shape of the outer envelopes of velocity outside of the ETM area are generally in agreement with the vertical distribution of open channel flow (*Figures 4-12a* and *b* are ebb-dominant, and *Figures 4-12e* and *f* are flood-dominant). However, at the core of the ETM in the middle reach

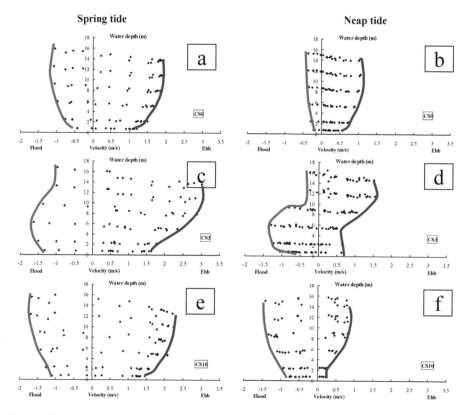

Figure 4-12.   Internal flow structures (the measured velocity against the corresponding water depth at each station during a tidal cycle) in the upper (CS0), middle (CS3) and lower (CS10) reach of the DNC during the spring (left panels) and neap (right panels) tide. Positive direction of velocity denotes ebb and negative is flood. The bold red curves are the outer envelopes of flood tide, the blue is ebb tide.

(*Figures 4-12c* and *d*), the vertical profiles deviate away from a classic logarithmic relationship. The lower half of the scattering appears to be concave in shape during the ebb-dominant period of both the neap and spring tides, whereas the near-bed part of the outer envelopes appears to follow the baroclinic pressure gradient during spring tide. In particular, the upper section (near-surface) is ebb-dominant but the lower section (near-bed) is flood-dominant during neap tide (*Figure 4-12d*).

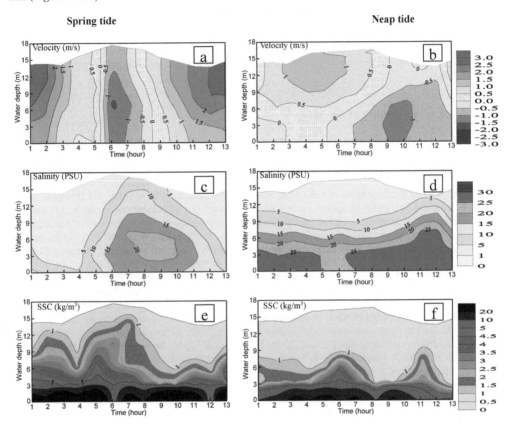

Figure 4-13. Time series of measured current velocity, salinity and SSC at station CS3, during spring (left panels) and neap (right panels) tides.

Figure 4-13 shows time series of measured current velocity, salinity and SSC at station CS3 during spring (left panels) and neap (right panels) tides. For most of the observation period (~10 h) during neap tide, the near-surface flow was directed down-estuary, but the near-bed currents were directed up-estuary. Salinity and SSC profiles are highly stratified, especially during neap tide (see *Figures 4-13d* and *f*). The vertical profiles of SSC and salinity are step-shaped, and a rather stable stratified density interface (see salt wedge (*Figure 4-13c*) and the 5 kg/m^3 contour lines *(Figure 4-13e* and *f)*) are formed clearly. The estimated gradient Richardson number can reach a value of more than 100 during slack water (see *Figure 4-5*). *Figure 4-13b* shows that the flooding process is confined by ebbing, and it appears to be originated from the bottom (in comparison with *Figures 4-13a* and *b*), indicating that the up-estuary baroclinic pressure gradient affects the vertical current profile. The reconstructed vertical velocity distribution further affects the patterns of residuals, controls the sediment and salt transport, and suppresses the intensity of near-bed turbulence. The effects of stratification and turbulence damping could be enhanced by each other and favor the formation of the up-estuary residuals. The process will pump enough sediment and salt mass to maintain a relatively stable ETM and salt wedge, and thus it is this mechanism that plays a dominant role in sediment trapping in the navigational channel.

In addition, *Figure 4-13* also shows the features of the vertical structures of currents during different tidal types. During the neap tide, at the core of the ETM in the middle reach (*Figures 4-13a and b*), the vertical profiles of velocity deviate away from a classic logarithmic form. Vertically, the maximum flood velocities during the neap and spring tides are not near the water surface. During the neap tide, the peak flood velocity is near the bottom and the flood period near the bottom layer is much longer than that near the surface. The lower half of the water column is clearly flood-dominant, while the upper section (near-surface) is ebb-dominant. It indicates that the vertical velocity structures are impacted by baroclinic pressure gradient.

4.4.3 Possibility of modeling

Modeling of observed estuarine phenomena offers the opportunity to improve our understanding of the physics of fine-sediment dynamics. Many previous studies *(Ross, 1988; Winterwerp, 1999; Le Hir and Cayocca, 2002; Guan et al., 2005; Cheng et al., 2009; de Nijs and Pietrzak, 2012)* have attempted to simulate the near-bed high concentration layer and reproduce the formation of the ETM, although few have tried to simulate the transport of fine sediment in a stably stratified water column. Through our aforementioned analyses, it has been shown that simulation of the transformed vertical velocity profile is crucial for modeling the dynamics of the ETM in a stably stratified estuary. Furthermore, major influencing factors that control the vertical velocity profile are the baroclinic effect and turbulence damping. Because density variations significantly alter fluid motion in stably stratified flows ($Ri_g > \sim 0.25–1$), the Boussinesq hypothesis (or Boussinesq approximation) is violated. Thus, the contributions of density differences to the inertial force term, the pressure differential term and viscous force term in the momentum equation cannot be ignored, and so some researchers (e.g. *Boonkasame and Milewski (2012)* have designated this non-Boussinesq flow. Laboratory and field data obtained in numerous studies *(Schumann and Gerz, 1995; Keller and Van Atta, 2000; Canuto et al., 2001; Sorbjan, 2006; Umlauf, 2009; Vasil'ev et al., 2011; Boonkasame and Milewski, 2012; de Nijs, 2012; Karimpour and Venayagamoorthy, 2014)* show that the critical feature of a stably stratified flow is a clear negative correlation between a decreasing inverse turbulent Prandtl number and an increasing gradient Richardson number. Consequently, successful simulation of the characteristics of stably stratified flows and associated sediment dynamics by a turbulence model requires this effect to be reproduced.

4.5 Conclusion

Spatial and temporal spring–neap variations of velocity, salinity and SSC in the DNC of the Yangtze Estuary during the 2012 wet season have been observed and analyzed in this study. From this observation data, the following conclusions can be made:

(1) With the channel narrowing and deepening, the North Passage, which is a primary navigational channel of the multi-channel estuarine system, is currently conditionally stably stratified, especially during periods of low tidal energies and slack waters.

(2) During a single tidal cycle, the longitudinal scale of the channel is sufficiently long so that the ETM cannot be displaced outside of its geographic boundary, and the convergent zone of residuals during both the neap and spring tides oscillates in the middle and lower reaches of the DNC. This encourages the formation of a near-bed high-SSC layer, which favors siltation in the dredged channel.

(3) Stratification and turbulence damping effects near to the ETM induce the upper half

(surface layer) of the estuary to be ebb-dominant and the lower part (near bed) to be flood-dominant, which is a previously undocumented phenomenon in this region.

(4) The residual pattern of currents, salt flux and sediment flux are non-similar in the stratified estuary, and the salinity-induced baroclinic pressure gradient is one of the dominant factors that control variation of vertical velocity structure. Additionally, residual analysis suggests that tidal resuspension and lateral sediment supply contribute to providing a sediment source during spring tides in the wet seasons.

(5) Field observations indicate that the transport of residuals generated by internal tidal asymmetry plays a dominant role in maintaining a stable-stratified density interface near to the ETM.

(6) Detailed investigation of the relationships between turbulence, the Prandtl number and the gradient Richardson number is needed for successful future modeling of stably stratified flows.

Experiment on settling velocity

Highlights

(1) A new insight determining the settling velocity of estuarine fine sediment is discussed in detail.

(2) An apparatus, approach and empirical formula are proposed for measuring setting velocity.

(3) We investigate the quantitative dependency of SSC, salinity and temperature on settling velocity.

(4) An empirical formula determining settling velocity of the Yangtze estuarine mud is proposed.

5.1 Introduction

For coastal sedimentology and morphodynamics, especially for those fine sediment estuaries with median grain size (D_{50}) finer than 62 μm, fall velocity (SV), also known as settling velocity, is a critical parameter in the understanding of sediment behavior and dynamics. SV directly determines the vertical distribution of suspended sediment concentration (SSC) and near-bed deposition flux, and its accurate determination has been regarded as a priority in characterizing fine sediment transport. In the past, extensive attempts, (e.g. *Rouse, 1938; McLaughlin, 1961; Owen, 1971; Mehta, 1989; Fennessy et al., 1994; Krishnappan, 2004; Gratiot et al., 2005; van Leussen, 2011*), have been made to delineate and decode the physical process of fine sediment settling and a number of empirical and semi-empirical formulas for SV have been proposed in various forms.

Generally, it is quite difficult to record or track the falling trajectory of a single fine-sediment particle or an individual floc visually or even using instrumentation. Three types of indirect methods have commonly been used to determine SV including (1) theoretical method (e.g., the Stokes formula, *Chien and Wan, 1999; Shao et al., 2011*) is used to calculate SV based on fluid properties (density and temperature) and grain size of an individual particle or floc. Moreover, some theoretical models by taking into account the behaviors of collision, coagulation and flocs growth and break-up have been setup to determine SV *(Winterwerp, 1998; Zhang and Zhang, 2011)*. (2) The settling velocity of fine sediment has been investigated using in situ method *(Owen, 1971; Fennessy et al., 1994; van Leussen, 1994; Agrawal and Pottsmith, 2000; Owen and Zozulya, 2000)* via an in situ settling tube or chamber. One of the disadvantages of the in situ method is the disturbance of the natural flocs during sampling and measurement. Moreover, gravity-induced settling is confounded by other physical processes, such as diffusion and advection, which maybe simultaneously occur. These processes cannot be removed or controlled during field surveys. Another insurmountable problem preventing in situ SV measurement is that this method cannot acquire a clear picture nor can it extract an individual floc under high SSC conditions. Slight increases in SSC can cause single flocs to become entirely indistinguishable, meaning that the method is only valid for relatively low SSC *(Markussen and Andersen, 2013)*. (3) Use of laboratory experiment (by settling column) for measuring the SV has a long history *(Camp, 1936; McLaughlin, 1961; van Leussen, 1988; Fathi-Moghadam et al., 2011)*. The main advantages of this method lie in the ability to measure SV under controllable conditions. Many factors, such as SSC, salinity and temperature, have been regarded as the independent variables of SV.

Based on previous studies on this issue, an experimental approach using an improved apparatus is presented for measuring the SV of suspended fine sediment in the laboratory. Subsequently, based on the experimental results, the dependencies of SV on SSC, salinity and temperature are tested and analyzed, and an empirical formula for determining SV is presented. The aim of this research is expected to implement those modeling so as to improve the internal structure of current and the stratification of SSC and salinity.

5.2 Method

5.2.1 Formula

The vertical mass balance equation has the following form if the diffusion term is assumed to be negligible:

$$\frac{\partial C}{\partial t} + \frac{\partial(\omega C)}{\partial z} = 0 \tag{5-1}$$

in which, C is the SSC, t is the time, z is the height above the bed and ω is the instantaneous depth-dependent SV at the depth h.

Integrating *equation 5-1* over depth yields

$$(\omega C)_{z=h} = -\frac{\partial}{\partial t}\int_0^h C dz \tag{5-2}$$

By multi-depth SSC sampling of a settling column at several time steps, the time- and depth-varied SV could be obtained using a finite-difference method (FDM). However, the problem is that SV is a parameter that is dependent upon other two independent variables (time and water depth), and is not practical for application. Therefore, a depth-dependent median SV is presented in *equation 5-2*, which refers to the concept of D_{50}. D_{50} is defined as the value of the particle diameter at 50% of the normal cumulative distribution *(Mehta, 2014)*, which is an easy and meaningful statistical way of quantifying the particle size of natural non-uniform sediment.

$$(\omega_{50\%})_{z=h} = \frac{1}{t_{50\%}}\int_0^{t_{50\%}} \omega_{i,h} dt \tag{5-3}$$

in which, $\omega_{50\%}$ is the median SV, $t_{50\%}$ is the elapsed time after the SSC decreased to 50% of its initial value at a given depth, and $\omega_{i,h}$ is the instantaneous depth-dependent SV.

The median SV means that when the SSC decreases by 50% over a certain depth, SV could be measured through the integration of the instantaneous SV over the elapsed time. Compared with other formulas, the McLaughlin formula has a strong physical meaning that is consistent with its original definition. It is widely employed for calculation of fine sediment SV *(e.g. McLaughlin, 1961; Tambo, 1964; You, 2004)*.

5.2.2   Experimental setup

5.2.2.1   Apparatus

From a series of experiments done in a traditional settling column with a diameter of 40 cm, a height of 2.5 m, and with eight multi-depth outlets for SSC sampling *(Wan, 2013)* *(Figure 5-1)*, the following drawbacks of traditional experiment procedures could be identified *(Berlamont et al., 1993; Dearnaley, 1996; Dyer et al., 1996)*: (1) Intakes are mounted at the fixed depths; therefore, the vertical profile of SSC is discontinuous to obtain the SV with a low resolution. (2) Owing to unfavorable wall effects during sampling, the sampled SSC would be smaller than the actual value, especially when the SSC value is higher than about 2-3 kg/m³. (3) When the SSC is lower than approximately 1 kg/m³, it is more prone to error induced by the drying weighting or pycnometric density methods. (4) Because sampling occurs at multiple depths and times, resulting in water volume loss the originally unrestricted settling process is disturbed (in still surroundings). (5) The traditional way of measuring/sampling the SSC is highly time-consuming; the average duration of one set experiment is about 4-5 days.

Figure 5-1. Photo of a traditional settling column, the outlets are connected together to a shaft to insure simultaneous sampling (Photo courtesy: *Huang Wei*).

A new apparatus for measuring both settling and consolidating velocities was designed and developed (*Figure 5-2*) to overcome those limitations above. The apparatus consisted of five parts: a settling column, an air agitator, an SSC monitor, a temperature control system and an electric lifting winch. The settling column is an acrylic and transparent cylinder with a diameter of 1 m and a height of 1.5 m, while the air agitator is driven by an electrical air compressor. Compressed air is guided by a circular acrylic board (with a diameter of 0.9 m) mounted horizontally 0.5 mm above the bottom of the settling column so that sufficient space remains to release the compressed air. Because compressed air was released near the bottom of the column and only at the edge of the board, SSC, temperature and salinity could be blended into a uniform suspension. The SSC monitor has two parts: a smart turbidity sensor (*Campbell Scientific Inc., OBS3+*) and a smart depth recorder (*RBR Ltd., DR-1050*). Furthermore, the temperature control system consists of eight 3-kilowatt temperature-controlled waterproof heaters and eight digital underwater thermometers. Heaters and thermometers are fixed on the inner wall at the bottom (i.e., about 0.5 m above the bottom of the column), and at the top (i.e., at about 0.5 m below the top of the column), respectively, in an alternating fashion and equally distanced from one another.

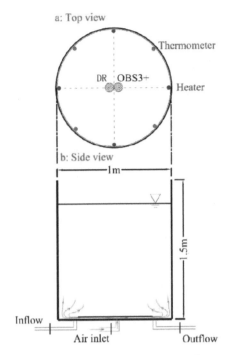

Figure 5-2. Schematic diagram showing the setup of the settling column. (a) Top view (the blue bold points denote heaters, and the red are thermometers), (b) side view and (c) photo of the settling column.

This apparatus has several advantages, including (1) the SSC value is continuously measured in the vertical direction by a smart turbidity sensor, which is more precise, more convenient and less time-consuming than the traditional way (drying weighting or pycnometric density method). (2) During the course of the experiment, almost no water loss and no external disturbances occur. (3) The chosen magnitude of the column diameter guarantees that the movement of the used turbidity sensor and depth recorder causes only very small disruptions of the settlement process. (4) Temperature, salinity, and initial concentration are easily adjustable. (5) The settling column can carry a tuning fork allowing for acoustic density measurement; thus it is suitable for self-weight consolidation experiments.

5.2.2.2   Procedure

The following are the main experimental procedures for measuring the SV:

(1) The water and sediment samples were pre-collected originally from the Yangtze Estuary (see *Figure 5-3* for the sampling location). The calibration relationship between SSC and turbidity of the OBS3+ sensor *(Sutherland et al., 2000)* is shown in *Figure 5-4*.

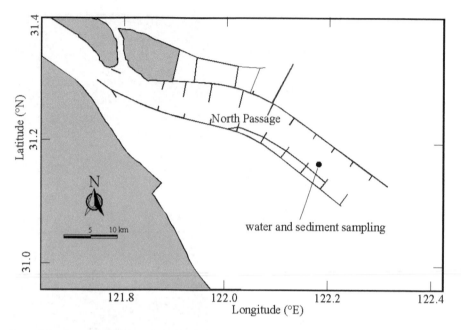

Figure 5-3. Locations of the water and sediment sampling.

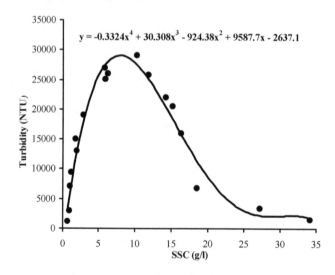

$$y = -0.3324x^4 + 30.308x^3 - 924.38x^2 + 9587.7x - 2637.1$$

Figure 5-4. Calibrated of SSC against the turbidity measured by using the OBS3+ sensor.

(2) To impound water sample into the settling column until the water depth is 1.2 m. Water samples are preferably gathered directly from in situ natural saline water, because the contents of the sample have a significant effect on the settling process of fine sediment. Use of artificial seawater therefore not recommended.

(3) To manually prepare the sediment sample using ~1000 ml natural seawater, to form a well mixed suspension. Put the sludge into the column and open the air agitator for 3 min so that proper blending occurs. Hold the column still for 1 min for adequate release of air bubbles and calming of the water surface, otherwise these bubbles may interfere with the signals of the

turbidity sensor. A preliminary measurement of SSC is conducted to check whether the value is as desired by lowering the turbidity sensor into the suspension with an electric winch. If the value is not in the desired range, this process is repeated by adding sludge and mixing the suspension. Impounding of sludge into the suspension step by step is crucial to successful blending. It should be noted that flocs are formed and grown up during the vertical settling or horizontal advection and diffusion processes in natural environment. Their existence is highly dependent upon local surroundings. They cannot be prefabricated and be picked up out of suspension independently. In our experiment, the aim of agitating the suspension (mixing) is not to "pre-produce" flocs, it just wants to break-up the consolidation status of sediment sample. Once those sediments are released in the settling column, flocs will develop naturally according to local physical and chemical characteristics of sediment and water samples.

(4) If the water temperature is lower than the required one, the heating system is turned on until all values indicated by thermometers are close to the required one. Then, the air agitator is switched on for 2-3 min to allow for uniform distribution of thermal energy and, consequently, thermal equalization throughout the settling column (i.e., removing temperature gradients).

(5) Once the temperature, salinity, and initial SSC are all close to the required values, agitation is ceased and the suspension is kept undisturbed for at least two min. The vertical distribution of SSC is measured to ensure its uniformity. The SSC difference between the top and bottom layer should be curbed to 5%.

(6) SSC is measured and recorded at relevant depths with the turbidity sensor and the depth recorder, which is connected to, and controlled automatically by, the electric lifting winch. The measurement interval is set to 5 min for the first 30 min and 10 min for the remainder of the experiment. Each set of experiments is completed when the value of the mid-depth SSC falls to 20% of the initial value.

(7) Based on time series of the vertical distribution of observed SSC, the median SV of the bottom layer is calculated using *equation 5-10*, and is taken as the representative SV for the experiment.

5.2.2.3   A pilot experiment

A pilot experiment was carried out to evaluate the impact of different water samples (i.e., in situ seawater, tap water and distilled water) on SV of fine sediment. Other factors that may influence SV were kept constant including the salinity of 0 psu, initial SSC of about 1 kg/m^3, the temperature of 20℃, and D_{50} of about 8 μm. The depth-dependent median SVs of the three different water samples are shown in *Figure 5-5*.

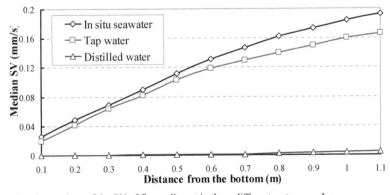

Figure 5-5.   Comparison of the SV of fine sediment in three different water samples.

As shown in *Figure 5-5*, the SV for distilled water was not of the same order as the SV for sea water or tap water, which is similar to the experimental results of *Wolanski (1995)*.

Furthermore, the SV for tap water is about 20% smaller than that for in situ seawater, indicating that the salt content makes an important contribution to the SV of fine sediment. Further investigation is needed to explain this phenomenon. Results of the pilot experiment indicate that seawater is best suited for this type of experiment.

5.2.2.4 Experimental setting

Consider both the practical demands of related research in the Yangtze Estuary, and the current understanding of the settling properties of fine sediment at this location, the scenarios of this experiment mainly focus on the impacts of SSC, salinity, and temperature on the SV of Yangtze estuarine suspended sediment. The experimental scenarios are shown in *Table 5-1*. The representative temperatures in the dry and wet seasons were calculated from the statistical measurement data for the Yangtze Estuary during 2006-2012. In addition, both sediment and water samples were collected from the Yangtze Estuary.

Table 5-1. Experimental scenarios for SV of the suspended sediment from the Yangtze Estuary.

Types	Temperature (℃)	Salinity (psu)	SSC (kg/m³)	Number of experiments
Dry season experiment	7	0, 4,7, 10, 15, 20	0.6-19	67
Wet season experiment	27	0, 5, 7, 9, 12, 15, 20	0.5-18	86
Temperature-varied experiment	7-30	7	1, 4.5	18

5.3 Result and discussion

It is generally understood that coarse-grained or higher density sediment particles have a higher SV. However, for fine particles, grain size is no longer a dominant factor. According to the literatures *(e.g. Guan, 2003; Winterwerp and van Kesteren, 2004; Mehta, 2014)*, there are a number of factors controlling fine sediment SV such as SSC, turbulence or turbulent shear, temperature, salinity, floc size and density, sediment composition, mineral and organic compositions, currents and wave climate, biological coatings, and the concentration of positive ions (such as Fe^{3+}, Ca^{2+}, Mg^{2+}, Na^+, K^+) in the suspension. In the following three subsections, the three most important factors are explicated to describe how SSC, salinity and temperature affect the SV of fine sediment.

5.3.1 Effect of SSC

SSC is regarded as the active scalar of SV and one of the major determinants of the fine sediment settling process *(Kineke and Sternberg, 1989; Gratiot et al., 2005; Manning et al., 2010)*. This dependence also constitutes a feedback loop as the vertical SSC distribution is itself the result of the settling process (the latter being the passive scalar). *Figure 5-6* shows schematized nonlinear relationships between SSC and SV, where two distinct stages are clearly depicted including accelerated flocculation settling and decelerated hindered settling. This means that, during the settling or deposition process, a small variation in SSC may cause great change in SV and sedimentation flux.

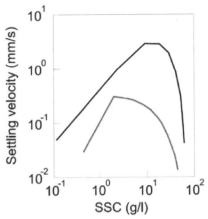

Figure 5-6. Schematic relationship between SSC and SV on a double logarithmic coordinate system. The black curve is redrawn from *Manning et al. (2010)*, while the red curve is redrawn from *Ross and Mehta (1989)*.

Using the improved apparatus and according to the median SV formula (*equation 5-2*), by plotting the relationship between SSC and median SV, the experimental SV values are shown in *Figure 5-7*. The magnitude of SV increases with increasing of SSC, until it reaches a threshold value after which it decreases with further increases of SSC. This phenomenon has been observed in other studies *(Mehta, 1989; Winterwerp and van Kesteren, 2004; Mikkelsen et al., 2007; Manning et al., 2010)*.

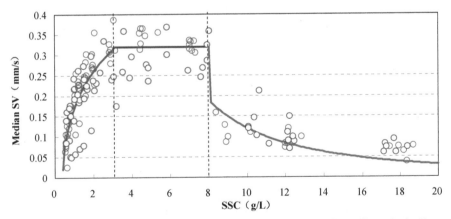

Figure 5-7. The relationship between SSC and median SV of suspended fine sediment in the Yangtze Estuary. The purple circles are SV in the wet season and the blue circles are in the dry season. Regarding the effect of SSC on SV, the settling process could be obviously divided into three stages: accelerated flocculation settling, maximum flocculation settling and decelerated hindered settling, which are depicted in the figure by purple, blue and red solid curves, respectively.

During the first stage, when SSC is in the range of 0.5-3 kg/m³, SV varies drastically as SSC changes. Thus, a small variation in SSC may trigger a rapid change in SV. The velocity is increased quickly by means of formation of larger and larger flocs with each step. The second stage is maximum flocculation settling, which has seldom been described by other researchers. In this stage, the SSC is in the range of 3-8 kg/m³. It seems that SSC has no influence on SV that it had earlier, and other factors such as salinity or temperature could replace its role. At this stage, with different surroundings, the size and density of a floc may reach a maximum and gravity may overcome its resistance and buoyancy. At the third stage, the flocculated network structure (flocs) may significantly control the settling process. The velocity is inhibited distinctly owing to the increase in size of flocs and decrease in density. Interactions among particles and flocs in the water column may reshape the settling behavior of fine sediment. The sensitivity of SV to SSC is hindered sharply during this stage.

The effects of SSC on SV of fine sediment have been discussed by many studies, although the degree is in a wide range. In particular, the critical SSC for maximum flocculation varies over a range of 1-30 kg/m³ *(Al Ani et al., 1991; Li and Mehta, 1998; Wu and Wang, 2004)*. Furthermore, the maximum median SV also varies over several orders of magnitude.

5.3.2   Effect of salinity

As shown in *Figure 5-8*, *Burt (1986)* has suggested that salinity could alter the SV of fine sediment in brackish and saline waters. Positive ions contained in natural seawaters, carry positive electric charges that benefit the growth of flocs. Higher salinity may induce greater SV over the entire SSC range according to *Burt (1986)*. However, higher saline content does not always favor SV; for example, SV reaches a maximum value at a salinity of 10-15 practical salinity units (psu) for Tamar Estuary mud *(Al Ani et al., 1991)*. In the Yangtze Estuary, this critical salinity for maximum flocculation is about 4-16 psu *(Jiang et al., 2002)*. Conversely, *Chen et al. (1994)* and *Berhane et al. (1997)* have argued that salinity has no significant effect on the SV.

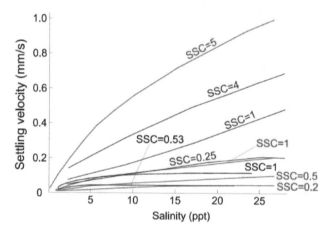

Figure 5-8. The relationship between SV and salinity, in the presence of different SSC conditions, which is summarized by *Burt (1986)*. This figure is redrawn after *Burt (1986)*. The unit of SSC is in kg/m³.

The sensitivity of SV to salinity is shown in *Figure 5-9*. Compared to SSC, salinity shows a subtle influence on SV. For each varied SSC condition (for both high and low temperatures), salinity exhibits a noticeable impact on the settling process of fine sediment. In the dry season, the critical salinity for maximum flocculation is about 7 psu. The effect of salinity is evident; the average ratio of the SV value under the condition of critical salinity to that under other salinity conditions varies in the range of 1.8-5.7. Similarly, in the wet season, the critical salinity value is around 10 psu. The salinity effect becomes weaker as temperature increases, and the average ratio of the SV under the condition of critical salinity to that under other salinity conditions varies in the range of 1.5-2.2. Overall, the effect of salinity on fine sediment SV should not be ignored and these significant impacts should be given more weight in future models of SV.

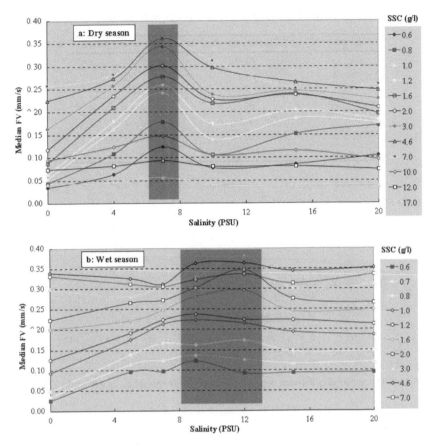

Figure 5-9. The relationship between salinity and median SV of suspended fine sediment in the Yangtze Estuary. The dark gray shaded zone refers to the maximum salinity for flocculation. (a) in the dry season (the water temperature is 7℃) and (b) in the wet season (27℃).

5.3.3   Effect of temperature

Temperature could also have an impact on the settling process of fine sediment. Many hydraulic engineers have already taken its effect on the kinematic viscosity (*Table 5-2*) in a water column, in accordance with the Stokes law *(e.g. Owen, 1972; van Rijn, 1993; Chien and Wan, 1999)*. Temperature could also affect the activity of individual particles and the motion of flocs. From a series of laboratory experiments, *Lau (1994)* concluded that the effective SV is higher for lower temperatures, which is exactly opposite to the results of *Owen (1972)*. Also from experimental data, *Jiang et al. (2002)* found that temperature is one of the most important factors affecting SV (*Figure 5-10*), and that a step-shaped relationship exists between temperature and SV. *Huang (1981)* demonstrated that temperature variation could lead to significant changes in SV (*Figure 5-11*).

Table 5-2. The standard relationships among temperature, density and kinematic viscosity *(Wu, 2007)*.

Temperature (°C)	Density (kg/m³)	kinematic viscosity (m²/s)
0	1000	1.79×10^{-6}
5	1000	1.51×10^{-6}
10	1000	1.31×10^{-6}
15	999	1.14×10^{-6}
20	998	1.00×10^{-6}
25	997	8.94×10^{-7}
30	996	8.00×10^{-7}

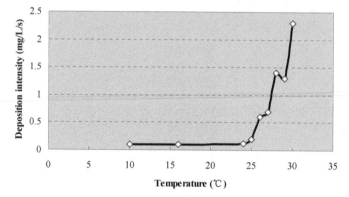

Figure 5-10. The relationship between deposition intensity and temperature, redrawn after *(Jiang et al., 2002)*.

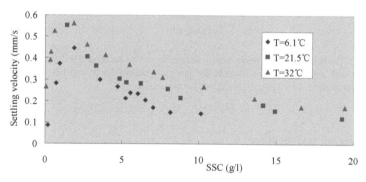

Figure 5-11. The relationship between SV and temperature, redrawn after *(Huang, 1981)*.

From *Figure 5-7*, it can be seen that SSC and SV are strongly correlated during the accelerating flocculation settling stage, but this correlation becomes weaker during the maximum flocculation settling stage. This indicates that other factors (e.g. temperature and salinity) more strongly influence the maximum flocculation settling stage relative to the accelerating flocculation settling stage. The effect of temperature on SV during these two settling stages was explored by conducting additional temperature-varied experiments (*Figure 5-12*). Overall, increasing temperature has a positive effect on SV. The influence of temperature varies with SSC as the variation of SV with higher SSC is much greater than that with lower SSC. Furthermore, when the SSC is over 8 kg/m³ the impact of temperature on SV is negligible (*Figure 5-7*).

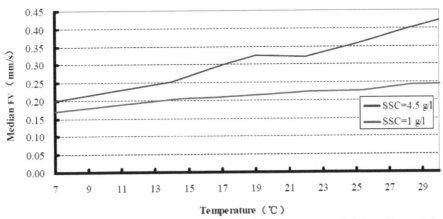

Figure 5-12. The relationship between temperature and median SV of suspended fine sediment. Salinity is 7 psu. Pink curve refers to the accelerating flocculation settling stage while dark curve to the maximum flocculation stage.

5.3.4 Formulation

Empirical formulas for SV of fine sediment can be found in the literature *(e.g. Ross and Mehta, 1989; Manning et al., 2010; Shi, 2010)*. In this study, a common expression is validated by regression analysis for different seasons (by varying temperature) to obtain an empirical formula for SV by using SSC and salinity.

$$\omega = (k_1(S - s_0)^2 + k_2) * C^{k_3} \qquad (0 \leq S < 30; 0 \leq C < 20) \tag{5-4}$$

in which, ω is SV (mm/s), S is salinity (psu), s_0 is the critical salinity for maximum flocculation (psu), C is SSC (kg/m³), and (k_1, k_2, k_3) are site-specified empirical fitting coefficients.

After fitting piecewise linear regression models to the data, the empirical coefficients for different seasons and different stages are summarized as below:

(1) For dry season
$$s_0 = 7$$

$$
\begin{cases}
\begin{cases} k_1 = -0.0067, k_2 = 0.22, k_3 = 0.49 & 4 \leq S \leq 10\,(a) \\ k_1 = 0.0005, k_2 = 0.10, k_3 = 0.41 & \text{other range}\,(b) \end{cases} \quad 0.5 < C \leq 3\ (1) \\[2ex]
k_1 = -0.0004, k_2 = 0.23, k_3 = 0.16 \qquad\qquad\qquad\quad 3 < C \leq 8\ (2) \\[2ex]
k_1 = 0, k_2 = 0.99, k_3 = -1.02 \qquad\qquad\qquad\qquad\quad 8 < C \leq 20\ (3)
\end{cases}
\tag{5-5}
$$

Note that the three stages (i.e., accelerating flocculation settling, maximum flocculation settling and decelerated hindered settling) are shown in (1) red, (2) blue and (3) purple, respectively. Case (a) indicates that the salinity condition is preferable for maximum flocculation; case (b) indicates salinity in other ranges.

(2) For wet season
$$s_0 = 10$$

$$\begin{cases} \begin{cases} k_1 = -0.0025, k_2 = 0.20, k_3 = 0.68 & 7 \le S \le 13\,(a) \\ k_1 = -0.0004, k_2 = 0.18, k_3 = 0.66 & other\ range\,(b) \end{cases} & 0.5 < C \le 3\ (1) \\[2mm] k_1 = -0.0001, k_2 = 0.41, k_3 = 0.12 & 3 < C \le 8\ (2) \\[2mm] k_1 = 0, k_2 = 0.99, k_3 = -1.02 & 8 < C \le 20\ (3) \end{cases} \qquad (5\text{-}6)$$

Correlations between experimental and empirical data for the dry and wet seasons are shown in *Figure 5-13*. It is clear that regression data from these empirical formulas are in good agreement with experimental data for both wet and dry seasons.

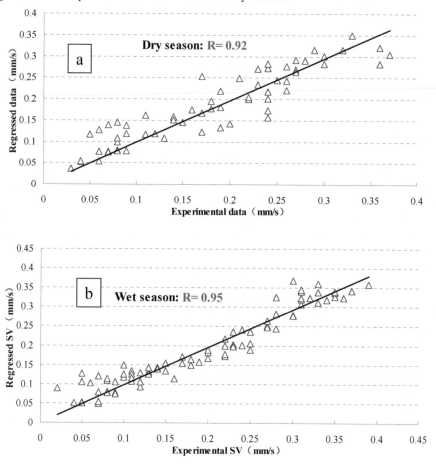

Figure 5-13.   Correlation between experimental and regressed SVs in the dry (a) and wet (b) seasons.

5.4   Conclusion

SV of fine sediment is a persistent issue that has attracted the attention of numerous scientists and innovative thinkers throughout the past several decades. Nevertheless, its underlying processes and mechanisms remain poorly understood, resulting in difficulty and confusion when engineers must select reasonable value for SV. In this study, we designed an improved apparatus to measure SV in the laboratory. Understanding and findings of the study are summarized in the following conclusions:

(1) Understanding settling characteristics of fine sediment is a complicated issue due to the cohesive property of particles. (i) During the settling process, the target (individual particle or flocs) which we want to measure its vertical moving speed is not fixed rigidly. Flocculation or de-flocculation will affect the size and density of the target sediment. That why direct measurement of fall velocity of fine sediment (record the variations of vertical position) is almost impossible. (ii) Not only the physical and chemical characteristics of sediment and water samples impact fall velocity of fine sediment, but also the hydrodynamic conditions of current has been thought could influence fall velocity. That is why in situ settling velocity measuring technique has appeared and developed in the past (it operated under "natural conditions"). In our opinion, the key point for those in situ methods is that they cannot separately distinguish the settling process from the vertical physical dynamic process. The variation of vertical position of sediment should not all attribute to the contribution of fall velocity. So, that is why we prefer to study the characteristics of fall velocity in the lab., not in situ. Owing to the limitations imposed on SV by in situ measurement methods, and the sensitive nature of the fine sediment settling process, laboratory experimentation is advocated for determining the fine sediment SV.

(2) SSC, salinity and temperature all affect SV, but to different extents. The relationships between SV of estuarine fine sediments and its various determinants (e.g., SSC, salinity and temperature) are highly dependent on specific environmental conditions. Furthermore, the impact of each determinant varies across the flocculation stages. During the first stage (accelerating flocculation settling), SSC is a dominant factor; conversely, in the second stage (maximum flocculation settling), dominant weights are held by salinity and temperature. And, more important, in the third stage (hindered settling), SV is least sensitive to all of these determinants compared to the first two stages.

(3) For the Yangtze estuarine mud, the SV peaks when SSC is in the range 3-8 kg/m^3, salinities for maximum flocculation settling are about 7 psu and 10 psu in the dry and wet seasons, respectively.

(4) Sediment research is highly empirically- and experimentally- oriented, with some key parameters highly site- and case-specific. Therefore, empirical formulas that rely on environmental factors are limited in their applicability to different field conditions.

(5) Compared to using a constant value or merely considering the limited effects of those SV determinants in numerical modeling, the proposed empirical formula could affect the vertical distribution of SSC directly and affect the overall sediment regime, sediment transport and morphodynamics of a fine sediment estuary, especially near the estuarine turbidity maxima zone.

On the internal current and SSC structures

Highlights

(1) Saltwater intrusion reshapes vertical structure of currents near an ETM area.

(2) Effect of flocculation settling velocity controls longitudinal ETM dynamics.

(3) Modeling of vertical profiles of current, salinity and SSC is dependent on the turbulent parameterization.

(4) The additional damping effect is functional to reproduce the SSC stratification in the DNC.

6.1 Introduction

Acting as a hub between inland and ocean, an estuary is amongst the most attractive places for humankind both ecologically and economically. Due to its high dynamics and variation, characterizing hydro- and morpho- dynamic processes in an estuary, especially within the turbidity plume zone, creates a great challenge to understand their responses to human activities and natural changes. The problem has been highlighted by the multi-disciplinary research community *(Corbett et al., 2004; Winterwerp and van Kesteren, 2004; Mehta, 2014)*. In an estuary system, the physical processes include: the balance between riverine freshwater inflow and marine saltwater intrusion, the interaction between tidal wave propagation and shallow water wave deformation, the competition between tidal mixing and density stratification, the presence of barotropic and baroclinic effects, and other processes generated by wind waves, Coriolis force (induced by earth rotation), sediment-current interaction, etc. contribute to the complexity of currents circulation and sediment dynamics.

In August 2012, a field campaign showed a distinct salinity and SSC stratifications in the North Passage (see *Figure 6-1*), one of the estuarine turbidity maximum (ETM) zones of the Yangtze Estuary *(wan et al., 2014a)*. Meanwhile, the vertical structure of currents was varying with the presence of the density stratification, and it even deviated from the law-of-the-wall *(King et al., 2008)*. The difficulty of simulating these phenomena (SSC and salinity stratifications, and variation of the vertical velocity profile) strongly motivates us to explore how these vertical structures are varied within the turbidity plume zone.

Within a turbidity plume or ETM zone, the vertical profile of suspended sediment concentration (SSC) is changed with the presence of near-bed high-concentration layer *(Winterwerp, 2006)*. Then, the existence of non-negligible vertical density difference in a partially mixed or stratified estuary is able to induce stratification effect on currents, especially close to the river bed. Furthermore, the density stratification affects the vertical turbulent eddy diffusion *(Wang, 2002)* and introduces a buoyancy effect *(Qiu et al., 1988)*. Both of them will substantially alter the vertical structures of current, salinity, SSC and turbulence *(Nepf and Geyer, 1996; Li et al., 2005; Burchard and Hetland, 2010; Cheng et al., 2011)*. In addition, because of the high dependency of SSC on settling velocity (SV) *(Manning et al., 2010; Mehta, 2014)*, the vertical distribution of SV is varied with the deformation of SSC vertical profile and the erosion and deposition fluxes are thus changed *(Krone, 1962; Partheniades, 1965)*. Most of these relevant understandings have been obtained qualitatively by numerical modeling e.g. *(Li and Zhong, 2009)* and field observation e.g. *(Uncles et al., 2006)*. It worth noting that, *Jay and Musiak (1996)* highlighted the effect of internal tidal asymmetry on the salt and sediment stratification in ETM area; and *Cheng et al. (2011)* compared the intensities of asymmetry tidal mixing (ATM) induced and density-induced flows in various types of density stratified estuaries. They emphasized the importance of ATM-induced flow to estuarine residual currents. *de Nijs and Pietrzak (2012)* demonstrated that the saltwater intrusion length is a critical parameter which governs sediment trapping in a stratified estuary. *Brors and Eidsvik (1992)* indicated that, for the simulation of turbidity currents, only the Reynolds stress model (RSM) appeared to be realistic, k-ε and algebraic stress models giving in particular wrong concentration profiles. *Murakami et al. (1996), Bardina et al. (1997)* and *Hattori et al (2006)* pointed out that the performance of the standard k-ε turbulence model is poor in simulation of a stably stratified flow; *Mellor (2001)* and *Guan et al.(2005)* stated that the Richardson-number-dependent turbulence dissipation is necessary in the modeling; many turbulence researchers *(Zilitinkevich et al., 2007; Vasil'ev et al., 2011; Karimpour and Venayagamoorthy, 2014)* proposed different empirical parameterizations for the turbulent Prandtl number to allow for the effect of density stratification; *Talke et al. (2009)* analyzed the separate influence from salinity structure, settling velocity and vertical mixing coefficient on the vertical SSC profile through an analytical model. A few studies *(Whitehead, 1987; van Rijn, 2007; DHI, 2009; Huang, 2010)* introduced a scaled (or named damping or stratification) factor to allow adjustment of the vertical turbulent diffusivity/viscosity for salt and sediment mixing. *van Maren et al. (2009)* compared the effect of various turbulent Prandtl numbers (0.7 and 2.0) on the vertical profile of SSC, based on the Delft3D model. All of these studies inspire us to do a systematic sensitivity analysis to enhance our understanding on fine sediment dynamics and to improve the modeling performance with respect to density stratification and its feedback on the velocity profile within a turbidity plume zone.

In this Chapter, firstly we report the observed time-series data which depicts the variation of vertical current SSC and salinity profiles in the ETM zone of the Yangtze Estuary. Subsequently, a schematized three-dimensional (3D) hydrodynamics, salinity and sediment transport model based on the Delft3D *(Deltares, 2014)* is built up. The objectives of the modeling are to reproduce and capture the observed phenomena with respect to SSC and salinity stratifications and variation of vertical velocity profile; and more importantly, to distinguish the dominant mechanisms and physical processes which are responsible for the phenomena through a series of sensitivity analysis tests and numerical experiments.

6.2   Methods

6.2.1   Field observation

Simultaneous measurements of current velocity, salinity and SSC were made hourly at three anchored stations (PT1, PT2 and PT3) on a neap tide (09:00 12 August - 10:00 13 August 2012) and spring tide (06:00 17 August - 07:00 18 August 2012), respectively *(Figure 6-1b)*. They were all deployed on the south side of DNC, 500m away from the center of the DNC). Data were collected at six relative water depth (0.05, 0.2, 0.4, 0.6, 0.8, and 0.95) *(Figure 6-2)*. PT2 is located at the core of the ETM zone in the North Passage *(Shi, 2010; Song and Wang, 2013; Wan et al., 2014b)*, whereas PT1 and PT3 are outside the ETM zone. The velocity data were obtained by rotor current meter. SSC and salinity were obtained after the water samplings were analyzed in the laboratory.

Figure 6-3 shows the vertical profiles of current velocity within the upper (PT1), middle (PT2) and lower (PT3) reaches of the DNC during the spring (left panels) and neap (right panels) tides. The envelopes of velocity outside of the ETM area comply with the law-of-the-wall (classic logarithmic relationship) *(King et al., 2008)* in general. *Figures 6-3-a* and *b* are ebb-dominant, and *Figures 6-3e* and *f* are flood-dominant. However, at the core of the ETM *(Figures 6-3c* and *d)*, the envelops apparently deviate the law-of-the-wall. The lower half of scatters appears to be concave in shape during the ebb-dominant period of both the neap and spring tides. In addition, it is should be noted that, the upper section (near surface) is ebb-dominant but the lower section (near bottom) is flood-dominant during neap tide *(Figure 6-3d)*.

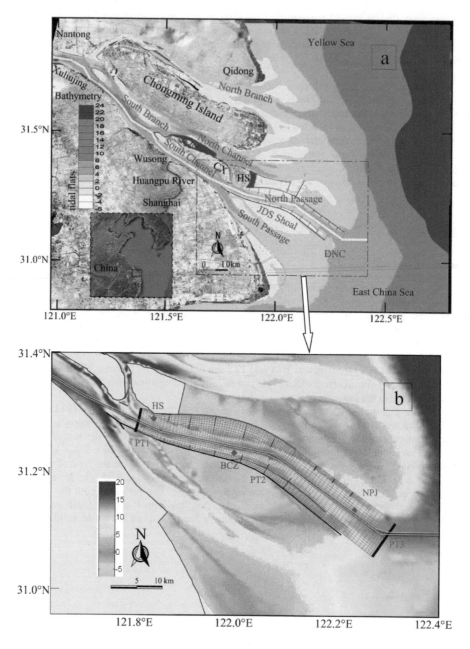

Figure 6-1. (a) General layout of the Yangtze Estuary. CX, HS, JDS, QR and DNS are the abbreviations of Changxing Island, Hengsha Island, Jiuduansha, Qingcaosha Reservoir and Deepwater Navigational Channel (DNC), respectively. The yellow bold line is the 12.5 m deepwater navigation channel; the purple lines denote the reclaimed lands; the black lines in the North Passage are the groins and dikes of the DNC project. (b) The model domain and horizontal mesh; the bold black lines are the open boundaries of the model. The red stars are the ship-born stations for currents, salinity and SSC; the purple diamonds denote the tidal gauging stations; the white lines mean the DNC. Note that the unit for the depth value is m.

Figure 6-4 exhibits the evolution of the vertical profiles of current velocity, salinity and SSC at station PT2 during the spring (left panels) and neap (right panels) tides. A distinct characteristic could be observed at the flood during the neap tide: the flow near surface was in ebb direction (*Figure 6-4b*, blue denotes flood, red means ebb), while that near bottom was in flood direction. The flood tide seems to be trapped and the internal flow structure is altered greatly. In addition, the current intensity seems too low to break the salinity stratification. During the spring tide, each high concentration event corresponds to a moment of peak velocity, suggesting tidal resuspension plays a quite significant role (see the pink dashed lines in *Figure 6-4*). The black dashed lines in *Figure 6-4* denote the strongest salinity stratification which takes places during the transition from flood to ebb tides.

Both salinity and suspension seem to be highly stratified, especially during the neap tide (see *Figure 6-4d* and *f*). Vertical profiles of SSC and salinity are L-shaped, and a rather stably-stratified density interface (see the 5 kg/m^3 contour lines in *Figures 6-4e* and *f*) and salt wedge (*Figure 6-4c*) are formed clearly. The estimated gradient Richardson number can reach several tens during the slack waters. The variations of internal structure of velocity, salinity and SSC imply that the residual currents, sediment transport, salt front, sedimentation and erosion within the turbidity zone are all impacted and varied.

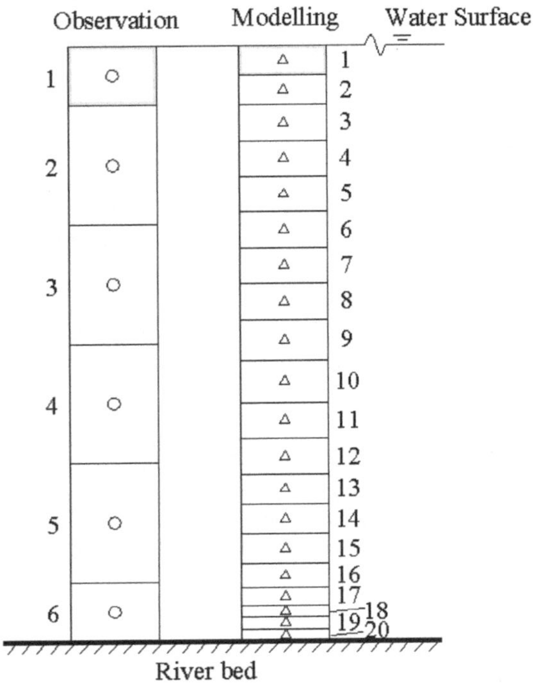

Figure 6-2. Comparison of the vertical grids between observation and modeling. The relative thicknesses of vertical layer for observation from surface to bottom are 10%, 20%, 20%, 20%, 20% and 10%, and the vertical layer thicknesses in the model are 5%, 5%, 6%, 6%, 6%, 6%, 6%, 6%, 7%, 7%, 6%, 6%, 5%, 5%, 5%, 4%, 3%, 2%, 2% and 2% from surface to bottom. The yellow lined grids are the examples of an individual layer. The monitoring points (marked by circle and triangle) are located at the center of each grid.

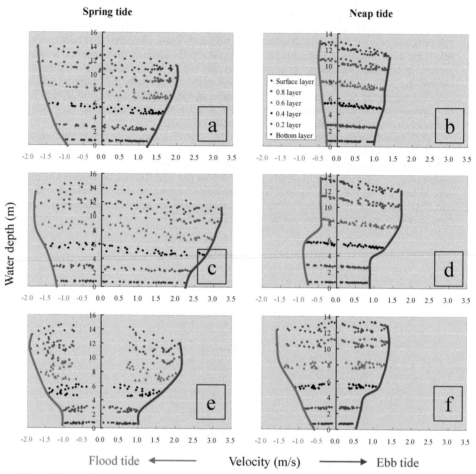

Figure 6-3. Vertical structures of flow (the measured velocity against the corresponding water depth at each station during a tidal cycle) in the upper (PT1), middle (PT2) and lower (PT3) reach of the DNC during the spring (left) and neap (right) tides. Positive direction of velocity denotes ebb and negative is flood. The bold red curves are the outer envelopes of flood tide, the blue is ebb tide.

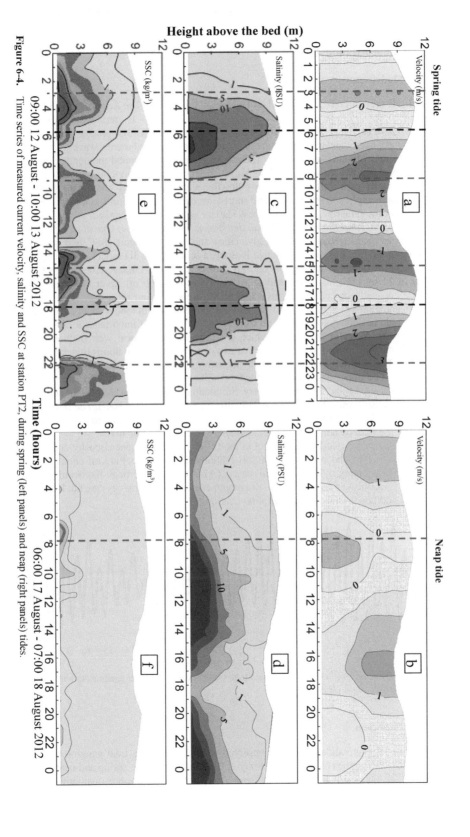

Figure 6-4. Time series of measured current velocity, salinity and SSC at station PT2, during spring (left panels) and neap (right panels) tides.

6.2.2 Numerical modeling

6.2.2.1 Model setup

Delft3D (version 6.01, developed by Deltares) *(Stelling and Van Kester, 1994; Stelling and Duinmeijer, 2003; Lesser et al., 2004; Deltares, 2014)* is an open source code, *(http://oss.deltares.nl/web/delft3d)* two- and three-dimensional (2D/3D) process-based *(Roelvink, 2006)*, hydrodynamic and transport numerical model for flows, sediments, waves, water quality, morphological evolutions and ecology. The model solves the horizontal momentum, continuity, mass transport and two-equation (such as k-ε) turbulence transport in a horizontally orthogonal curvilinear mesh with vertically terrain-following sigma or z-level coordinate system. The model has been applied to study long- and short-term current circulation and transport processes in many riverine, estuarine and coastal areas *(Lesser et al., 2004; Dastgheib et al., 2008; Hu et al., 2009b; de Nijs and Pietrzak, 2012; van der Wegen and Jaffe, 2014; Guo et al., 2014).*

To better characterizing the hydrodynamics and mass transport (SSC and salinity) just within the turbidity plume of the Yangtze Estuary, a schematized 3D model was set up. For the sake of simplification the model domain only covers the North Passage (see *Figure 6-1b*) and consists of 25 by 111 grids (the smallest grid cell size is ca. 60 m). Over the depth 20 sigma layers are unequally spaced, the relative thicknesses are shown in *Figure 6-2*. The elevations of the jetties and groins (see *Figure 6-1*) for the DNC project are 2 m (above local mean sea level), and they could be submerged at the high waters. Therefore, it is of importance to allow for these engineering structures in the model, which are defined by the local weir method *(Deltares, 2014)*. The model is forced by water levels at two open boundary conditions (black bold lines in *Figure 6-1b*). The water levels are given by the nearest tidal gauging stations HS and NPJ (see *Figure 6-5*). The upper and lower boundary conditions of SSC and salinity are provided by measured data at PT1 and PT3, respectively. Simulations start at 0h on August 10 and end at 0h on August 19, 2012, modeling starts with an initial condition file (hot start). The spin-up time of modeling is 10 days from 1 to 10 August 2012. One fraction of cohesive sediment is used in the model. The physical parameters of the modeling are referred to *Table 6-1*.

In the standard k-ε turbulence model in the Delft3D, the turbulent viscosity and diffusivity for both momentum and materials transport (salinity and sediment) are all equal if no wave action. For the damping experimental cases (A7-A11), the vertical momentum diffusion coefficient (viscosity) and vertical mass dispersion coefficient (diffusivity) are not kept the same to drive an additional effect in the transport equations.

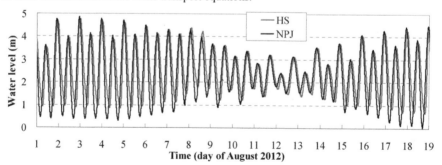

Figure 6-5. Time series of (hourly) measured tidal water levels at the tidal gauging stations HS and NPJ (see *Figure 6-1b* for their positions).

6.2.2.2 Validation

The model was validated against observed water level at the tidal gauging station BCZ (*Figure 6-6*). Time-series of SSC and salinity at PT2 during both the spring and neap tides were validated, see the result of Case 11 *(Figure 6-11)*. Model results are qualitatively in good agreement with measured ones. These main characteristics, such as salinity stratification, internal current structure deformation, sediment resuspension and trapping are well captured by the schematized model.

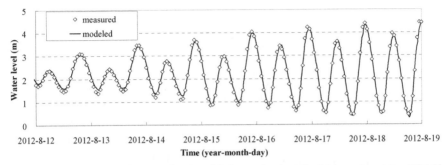

Figure 6-6. Comparison of measured and modeled water levels at the tidal gauging station BCZ (time intervals of measured and modeled results are 1 h and 10 min, respectively).

6.2.2.3   Numerical experimental setting

Because of our poor understanding of sediment transport, many physical parameters are "free" (they are of wide range and some physical processes could be expressed in various ways) in the modeling. In different case studies or even in the same study area, one could reasonably calibrate the model with different parameters. Sensitivity analysis is an essential and effective way to gain insight into the physical processes. In this section, 13 numerical experiments *(see Table 6-1)* were undertaken to examine the factors which control the vertical structures of current, salinity and SSC in the North Passage. The objective of Cases A0-A3 is to investigate baroclinic effects on vertical structure of current. Cases A3-A6 are used to explore the sensitivity of settling velocity to vertical SSC profile. Cases A3, A7 and A8 are built to compare the performance of different turbulent models. Cases A3, A9-A11 are conducted to understand the buoyancy effect and additional Ri-dependent turbulence damping effect. Case 12 is designed to observe the effect of drag reduction on SSC profile. The sensitivity analyses are carried out during both the spring and neap tides.

Table 6-1. List of configurations of the numerical experiments.

Case No.	Specified settings	Common setting
A0	Without sediment and salinity transport (Hydrodynamic model)	
A1	Only salinity, but without sediment transport	
A2	With sediment and salinity transport both, but without effect of sediment on fluid density	
A3 (reference case)	Sediment and salinity buoyancy effect included with effect of sediment on fluid density Standard k-ε turbulence model With the experimental Yangtze mud fall velocity	Time step: 0.5 min
A4	Constant fall velocity 0.04 mm/s	Manning coefficient: 0.011
A5	Constant fall velocity 0.4 mm/s	Horizontal viscosity and diffusivity: 0.1
A6	Constant fall velocity 4 mm/s	Background vertical viscosity and diffusivity: 1e-5
A7	ALG turbulence model (Ri dependent)	Critical erosion shear stress: 0.1 N/m²
A8	K-L turbulence model (Ri dependent)	Critical deposition stress: 1000 N/m²
A9	scaled k-ε model, viscosity and diffusivity are damped simultaneously based on Ri (vertical averaged Ri)	Erosion rate: 2e-4
A10	scaled k-ε model, viscosity and diffusivity are amplified with a constant damping factor (1.1)	
A11 (validation case)	scaled k-ε model, only viscosity is amplified based on Ri (vertical averaged Ri)	
A12	Consider the drag reduction effect based on Case A3 (See *equation 6-9*)	

6.3 Result and discussion

6.3.1 Effect of baroclinicity

With the presence of dissolved and suspended masses or due to the non-uniform distribution of temperature in water column, the fluid density would be altered introducing significant variation on pressure term of the momentum equation. According to the pressure *equation 6-1*, the baroclinic effect is directly associated with the longitudinal density gradient, which implies that, in principle, only vertical density gradient would not cause the baroclinic effect. In many shallow estuarine systems, saltwater is considered as a primary contributor to the effect of baroclinicity *(Geyer and Farmer, 1989; Uncles et al., 1990; Burchard and Baumert, 1998; Nguyen, 2008; de Nijs and Pietrzak, 2012)*, due to the fact that the magnitude of salinity is one or two orders greater than SSC. For example, in the Rotterdam Port, SSC is about 0.01-0.3 kg/m³, but accordingly salinity reaches around 1-30 psu *(de Nijs and Pietrzak, 2012)*. Following the water density *equation 6-2 (Guan et al., 2005)*, the weight of salinity is much more important than the concentration in this condition. Once the magnitude of SSC becomes comparable to salinity, its contribution to the fluid density should be taken into account.

$$\frac{1}{\rho_0}\frac{\partial p}{\partial x} = g\frac{\partial \xi}{\partial x} + \frac{g}{\rho_0}[\int_z^\xi \frac{\partial \rho}{\partial x}dz'] \tag{6-1}$$

$$\rho \approx 1000 + 0.78S + 0.62C \tag{6-2}$$

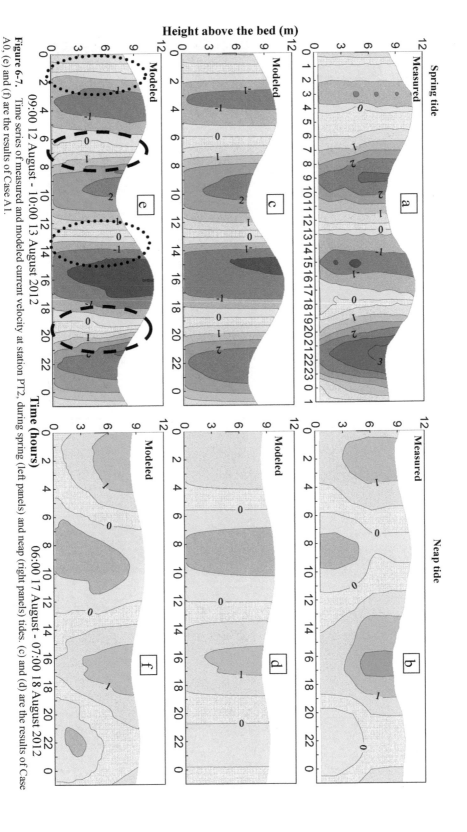

Figure 6-7.   Time series of measured and modeled current velocity at station PT2, during spring (left panels) and neap (right panels) tides. (c) and (d) are the results of Case A0, (e) and (f) are the results of Case A1.

Multiscale physical processes of fine sediment in an estuary

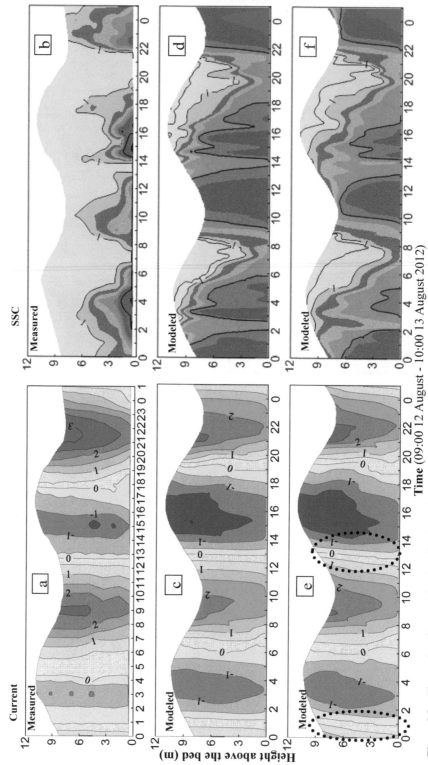

Figure 6-8. Time series of measured and modeled current velocity (left panels) and SSC (right panels) at station PT2 in spring tide. (c) and (d) are the results of Case A2, (e) and (f) are the results of Case A3.

where ρ is the fluid density in kg/m³, S is salinity in psu and C is SSC in kg/m³.

Figure 6-7 shows modeled internal current structures with and without salinity as well as measured ones. During the neap tide, measured results show a typical phenomenon "flooding tidal trapping", i.e. the core of flooding tides is restricted in the lower part of the water column (*Figure 6-7b*). It could be observed that the typical phenomenon is only captured in the case with salinity (*Figure 6-7f*), but not by that without salinity (*Figure 6-7d*). Because tidal energy during spring tide is much stronger than that during neap tide, the velocity profile seems to be less impacted by the baroclinicity. However, when the current intensity is relatively weak as marked by the elliptical dash lines in *Figure 6-7e*, the gradients of the velocity isolines are changed a little in comparison to those in the case without saltwater (*Figure 6-7c*). In addition, it is worth noting that the flow structures at the time point 1 and 13 as denoted by dotted ellipse lines in *Figure 6-7e*, seem to be less impacted due to the fact that the salinities at these time points are much less than that at the times 7 and 19 (see *Figure 6-9*). In sum, the flow structure is altered by saltwater intrusion induced baroclinic effect, not only during neap tide but also during spring tide; the effect becomes more evident at lower currents.

Cases A2 and A3 are designed to examine the effect of sediment concentration on fluid density. On the one hand, as aforementioned, while the range of SSC is in one or two orders smaller than salinity value, the SPM-induced baroclinic effect is of secondary importance. Nevertheless, the effect of SSC on fluid density could be identified from the comparison of current structure between *Figures 6-8c* and *e*, especially at the moments of low currents, which are indicated by the dotted ellipse. Comparison between *Figures 6-8a* and *e*, especially in the area marked by dotted ellipse, the simulated result with the effect of sediment on fluid density seems more reasonable than that without. On the other hand, the effect of SSC is able to further affect local SSC profiles (*Figures 6-8b, d* and *f*). Even though there is a substantial difference between the modeled and observed SSC profiles; it could be still found that the vertical distribution of SSC is more reasonable when the effect of sediment on fluid mud is taken into account. Due to this effect, more sediments could be trapped at the location with high turbidity.

The group of sensitivity tests addresses the important role of saltwater- and sediment-induced baroclinic effect on the internal flow and SSC structures. Saltwater intrusion strongly controls vertical structure of currents near the estuarine turbidity maximum zone, especially at lower currents, during which the baroclinic pressure gradient forcing could significantly reshape the local velocity profile. Meanwhile, the effect of SSC on fluid mud density is also found to favor the stratification of SSC.

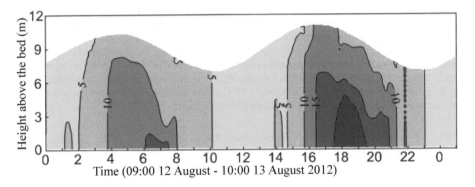

Figure 6-9.   Time-series of modeled salinity profile of the Case A1.

6.3.2   Effect of settling velocity

Settling velocity of fine sediment in estuaries has received attention from both coastal scientists and hydraulic engineers *(van Leussen, 1988; Eisma et al., 1996; McAnally, 2000; Guan, 2003; Mantovanelli, 2005; Pejrup and Mikkelsen, 2010)*. It has long been realized that for coarse-grained sediments, denser and larger particles have a higher SV, but for fine particles, the individual (single) grain size is not a dominant factor any more. According to the literature *(Han and Lu, 1983; Dearnaley, 1996; Guan, 2003; Winterwerp and van Kesteren, 2004; Mehta, 2014)*, there are a number of factors which can influence the SV of fine sediment, such as SSC, salinity, temperature, water depth, turbulent intensity, floc size and density, sediment ingredients, mineral and organic compositions, and even biological coatings in the suspension.

As shown in *Chapter 5*, based on the experimental results in the laboratory, an empirical formula, *equation 5-4*, has been proposed with caution to approximately determine SV of the Yangtze estuarine mud. The influences from SSC, salinity and temperature on SV are all considered in the experimental-oriented formula. The formulation can well capture the flocculation-induced acceleration and hindered settling processes.

In order to investigate the feedback of settling velocity on fine sediment dynamics, we carried out 4 experimental case studies (see *Table 6-1*, Cases A3-A6). Settling velocity in Case A3 varies with SSC and salinity. Empirical coefficients are selected for the wet season (see *Section 5.3.4*). SV in Cases A4 is 0.04 mm/s, in Case A5 0.4 mm/s, and in Case A6 4 mm/s.

Jung et al. (2004) and *Talke et al. (2009)* derive the concentration distribution equation from a one-dimensional vertical advection–diffusion equation (*equation 6-3*). It is suggested that the suspended sediment stratification is largely determined by the Peclet number. Larger pe_v indicates strong sediment stratification. Thus, the sediment stratification can be interpreted from SV, water depth, and smaller mass diffusion.

$$\frac{C_z}{C_b} = \exp\left\{-pe_v(\frac{z+H}{H})\right\} \qquad (6-3)$$

where, C_z is the SSC at the height of z, z is the height above the bottom, C_b is the SSC at the bottom, $pe_v = \frac{\omega_s H}{K_v}$ is the Peclet number, K_v is the vertical eddy diffusivity coefficient, ω_s is the settling velocity and H is the water depth.

Since the settling velocity is directly associated with the vertical distribution of SSC, in this section we mainly focus on the effect of SV on SSC. *Figure 6-10* shows time series of modeled SSC in the different cases. In Cases A4-A6, with increasing SV, SPM remained in the water column decreases and SSC reduces accordingly. The mass of suspended sediment seems to be sensitive to the variation of SV and actually is also determined by deposition flux at the bottom layer. The deposition flux (*equation 6-4*) depends on the product of sedimentation efficiency, fall velocity and near bed concentration when bottom shear stress is smaller than a threshold value *(Partheniades, 1965; van Kessel et al., 2011b; Deltares, 2014)*. Therefore, the SSC in the water column could be increased by decreasing the deposition flux. In this study, the deposition flux is independent of the critical stress for deposition which is equal to 1000 Pa in modeling as suggested by *Winterwerp and van Kesteren (2004)*. Thus the SSC in the water column can be only increased through reducing SV or sedimentation efficiency.

$$D = \alpha \omega_s C \qquad (6-4)$$

where, ω_s is the SV, α ($0 \le \alpha \le 1$) is the sedimentation efficiency, C is the SSC, and D is the bottom deposition flux.

Comparisons of modeled results in *Figures 6-10c, d* and *Figures 6-10g, h* with the measured ones show that modeled results of Cases A3, A5 reproduce the four high SSC events around the times 4, 11, 15 and 0 during the spring tide and around the time 6 during the neap tide, which agrees with the measured ones. In view of the intensity of stratification, Case A3 is even better, with sharper stratification than Case A5.

According to *equation 6-3*, when SSC is larger than 8 kg/m³, the hindered settling occurs. However, the bottom SSC in *Figure 6-10c* is only 3-4 kg/m³, which means that the bottom deposition flux is not affected by the hindered settling. Once the SV at the bottom is reduced due to the hindered settling, SSC near bed would increase according to comparisons among Cases A4-A6. That implies the hindered settling also affects SSC stratification. Therefore we have to keep in our minds that the hindered settling effect is only activated when SSC in modeling is high enough to reach the gelling point *(Winterwerp, 1999)*. Otherwise, even the hindered settling is defined in the numerical model, but never really functions just like Case A3 does.

Another effect of the hindered settling is to avoid "sudden-death" for the near bed sediments. According to the definition of the deposition flux term, the "switch" can be turned on once the bottom shear stress is less than a threshold value. Then the sediment is removed from the water system immediately. A substantial gap is remained between our understanding on the consolidation process and the numerical process *(Taylor and Leonard, 1990; Le Hir et al., 2011; Slaa et al., 2013)*. Due to the hindered settling and consolidation process are much slower than flocculation settling process. The hindered settling is able to buffer the "sudden-death" process moderately by decreasing the near bed settling velocity, and then more sediment will be released into the system. In short, the hindered settling makes the near bed sedimentation and consolidation processes more reasonable than without the effect in those high turbidity environments.

The settling velocity is a function of salinity (see *equation 6-5*) and SSC (hindered settling, see *equation 6-6*), and defined in the standard Delft3D model *(Deltares, 2014)*. As shown in *Figures 6-11*, only salinity is able to raise SV, but SSC not. In addition, effect of SSC on SV seems to be quite limited during the deceleration stage, and effect of the hindered settling is relatively weak.

$$\omega_s = \begin{cases} 0.5\omega_{s,max}(1-\cos(\dfrac{\pi S}{S_{max}}))+0.5\omega_{s,f}(1+\cos(\dfrac{\pi S}{S_{max}})) & (S \leq S_{max}) \\ \omega_{s,max} & (S > S_{max}) \end{cases} \tag{6-5}$$

where, $\omega_{s,max}$ is the maximum settling velocity when the salinity is greater than S_{max}, S_{max} is the threshold salinity, $\omega_{s,f}$ is the settling velocity in freshwater, S is the salinity. Herewith we take $\omega_{s,f} = 0.05 \ mm/s$, $\omega_{s,max} = 0.3 \ mm/s$ and $S_{max} = 10$ psu for an example.

$$\omega_{s,h} = (1-\dfrac{C}{C_r})^5 \omega_{s,0} \tag{6-6}$$

where, $\omega_{s,h}$ is the hindered settling velocity, C is the SSC, C_r is the reference density, $\omega_{s,0}$ is the settling velocity without hindered effect but the effect of salinity is included.

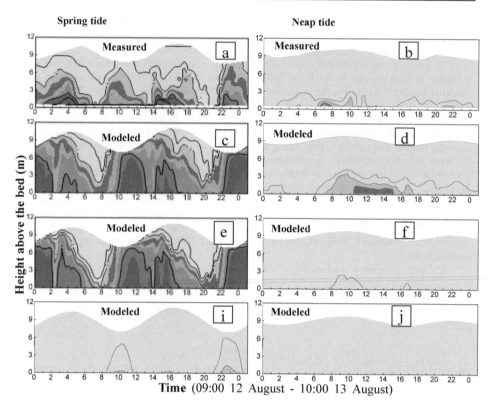

Figure 6-10. Time series of measured and modeled SSC at station PT2, during spring (left panels) and neap (right panels) tides. (c) and (d) are the modeled results of Case A3, (e) and (f) are the results of Case A4, (g) and (h) are the results of Case A5, (i) and (j) are the results of Case A6.

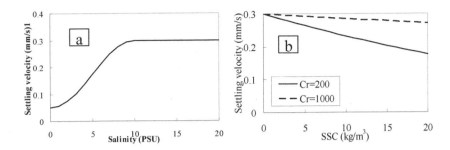

Figure 6-11. The dependency of settling velocity on salinity (a) and SSC (b). $\omega_{s,0}$ =0.3 mm in (b) is an example to show effect of the hindered settling, actually it should be varied with salinity.

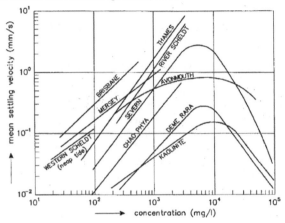

Figure 6-12. Empirical relationship of SV dependence on SSC *(van Rijn, 1993)*.

6.3.3   Effect of turbulence damping

Flows laden with salinity or suspended sediment are ubiquitous in nature. With the salt mass and SPM a high anisotropy and inhomogeneity are introduced into the turbulent flow, especially near the river bed. In a stratified water column, the strong vertical density gradient could have significant buoyancy effects on the turbulence *(Jones, 1973; Pacanowski and Philander, 1981; Nunes Vaz and Simpson, 1994)*. The buoyancy-induced turbulence is given by,

$$B_k = \frac{A_v g}{\rho H \, \mathrm{Pr}_t} \frac{\partial \rho}{\partial z} \tag{6-7}$$

where, B_k is the buoyancy flux, A_v is the vertical eddy viscosity, Pr_t is the Prandtl-Schmidt number.

In case of strong vertical density gradients, the turbulent exchanges are considerably inhibited by buoyancy forces. So the flow turbulence in a stratified water column is significantly suppressed *(Winterwerp, 2011b)*, and subsequently the diffusion of materials (SPM and salinity) is also impacted. In order to model vertical profiles of SSC correctly, it is of great importance to understand turbulence damping effects in the stratified estuary *(Whitehead, 1987; van Rijn, 2007; DHI, 2009; Huang, 2010)*. Originally, if one of those two-order turbulence models *(Umlauf and Burchard, 2005; Warner et al., 2005)* is selected and the baroclinic effect is considered in a hydrodynamic model, the vertical density-induced turbulence damping is produced (see *Figure 6-13*).

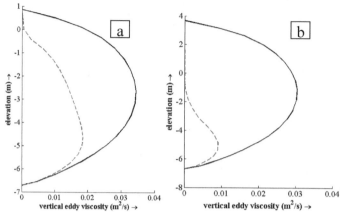

Figure 6-13. Comparisons of the vertical eddy viscosity without (black lines, Case 0) and with (red dashed lines, Case 3) baroclinic effect at station PT2 at the peak ebb (a) and peak flood (b) times.

In Case A3, although turbulence damping has been included (*Figure 6-13*), there is a discrepancy between modeled vertical profiles of SSC (*Figure 6-10c*) and measured ones (*Figure 6-10a*). To further improve the vertical profile of SSC in *Figure 6-10c*, additional (enhanced) turbulence damping has to be taken into account. It is normally involved into modeling through a widely-used parameterization between the turbulent Prandtl number (Pr_t) and the gradient Richardson number (Ri_g) (*Abarbanel et al., 1984; Peters et al., 1988; Rodi, 1993; Zilitinkevich and Esau, 2007; Geyer et al., 2008; Venayagamoorthy and Stretch, 2010*). Some of these empirical formulae have been listed in *Table 6-2*, the relationship between Pr_t and Ri_g is shown in *Figure 6-14*.

Table 6-2. Empirical formula parameterizating the relationship between the turbulent Prandtl number (Pr_t) and the gradient Richardson number (Ri_g).

Formula	Reference and its abbreviation	Parameter
$Pr_t = \dfrac{(1+3.33Ri_g)^{1.5}}{\sqrt{1+10Ri_g}} Pr_{t0}$	*(Munk and Anderson, 1948)* MA	$Pr_{t0} = 0.5 - 1.0$ *(Kays, 1994)*
$Pr_t = \dfrac{\dfrac{A_0}{(1+5Ri_g)^2}+A_b}{\dfrac{A_v}{1+5Ri_g}+K_b}$	*(Pacanowski and Philander, 1981)* PP	$A_v = \dfrac{A_0}{(1+5Ri_g)^2}+A_b$ $A_0 = 5\times10^{-3}$, $A_b = 10^{-4}$, $k_b = 10^{-5}$
$Pr_t = (1+3Ri_g)^2$	*(Lehfeldt and Bloss, 1988)* LB	
$Pr_t = \dfrac{5(1+5Ri_g)^{-1.5}+0.2}{5(1+5Ri_g)^{-2.5}+0.01}$	*(Strang and Fernando, 2001)* SF	
$Pr_t = Pr_{t0}+5Ri_g$	*(Zilitinkevich et al., 2007)* ZE	
$Pr_t = Pr_{t0}\exp(-\dfrac{Ri_g}{Pr_{t0}\Gamma_\infty})+\dfrac{Ri_g}{R_{f\infty}}$	*(Venayagamoorthy and Stretch, 2010)* VS	$\Gamma_\infty \approx 1/3$, $R_{f\infty} \approx 0.25$
$Pr_t / Pr_{t0} = (1-\dfrac{z}{D})\dfrac{Ri_g}{R_f}+Pr_{tw0}$	*(Karimpour and Venayagamoorthy, 2014)* KV	$Pr_{tw0} \approx 1.1$, $R_f \approx 0.25$ when $Ri_g > 0.25 - 0.4$

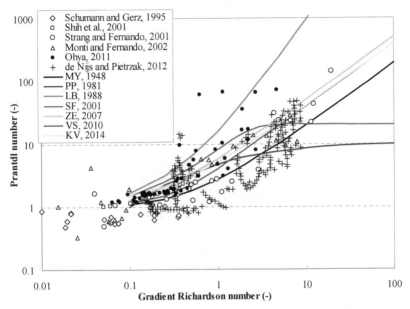

Figure 6-14. The relationship between Ri_g and Pr_t. (the various types of scatter points are measured or experimented data *(Schumann and Gerz, 1995; Ohya et al., 1997; Strang and Fernando, 2001; Monti et al., 2002; de Nijs and Pietrzak, 2012)*, and the lines are empirical parameterizations, the abbreviations are referred *Table 6-2.*)

It is worth noting that, in the modeling, Ri_g is explicitly calculated by a division (*equation 6-8*). In numerical computation we have to be always very careful about those divisional or fractional expressions, because a small error from the denominator may propagate fast, be enlarged continuously and finally become out of control. This numerical instability was once encountered in the study of *Huang (2010)*. *Dyer and New (1986)* highlighted that both vertical interval and velocity precision affects the magnitude of Ri_g. *de Stadler et al. (2010)* pointed out that a larger Prandtl number (higher level of stratification) can significantly result in a higher computational cost. *Figures 6-15* and *6-16* demonstrate that calculated Ri_g is oscillated temporarily and spatially. Such a kind of computational instability would further impact eddy viscosity and diffusivity via the parameter Pr_t. To increase the precision of Ri_g, the horizontal and vertical mesh resolutions have to be both improved, but a much higher computational cost is needed. The same numerical instability is also found in the Cases A7 and A8 (*Figure 6-17*), because both the algebraic and K-L turbulence models introduce a Ri_g-dependent damping parameter to account for the effect of density stratification *(Deltares, 2014)*.

$$Ri_g = \frac{g\frac{\partial \rho}{\partial z}}{\rho(\frac{\partial u}{\partial z})^2} \qquad (6\text{-}8)$$

where g is gravitational acceleration, ρ is density of saltwater containing suspended sediments, $\frac{\partial \rho}{\partial z}$ is the vertical density gradient, $\frac{\partial u}{\partial z}$ is the vertical gradient of horizontal velocity.

The mass transport is strongly dependent on a correct prediction of the development of turbulence. For the fine sediment, they are coupled and interacted with each other. To reduce the differences between model prediction and observation for SSC profiles, a better description of vertical mixing is needed, but could not be simply achieved by tuning vertical eddy viscosity and diffusivity *(Huang, 2010)*. Nevertheless, turbulent viscosity and diffusivity are of primary importance to our understanding of how large the turbulence damping affects model prediction of stratified flow. Thus, another sensitivity study (Cases A9-A11) was conducted with three types of scaled formulae, in which the enhanced turbulence damping effect is taken into account. In Cases A9 and A10, a $\overline{Ri_g}$-dependent scaled parameter is introduced, $\overline{Ri_g}$ is the vertical averaged value to smooth the fluctuating data to some extent; and in Case A11 a constant scaled stratification factor is used to account for the enhanced damping effect.

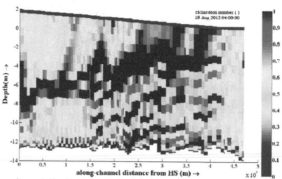

Figure 6-15. Along-channel distribution of calculated Ri_g at the peak ebb of spring tide of the Case A3.

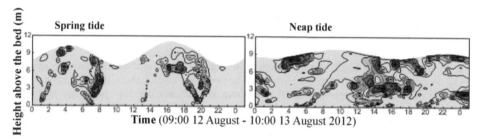

Figure 6-16. Time-series of calculated Ri_g during the spring and neap tides of the Case A3.

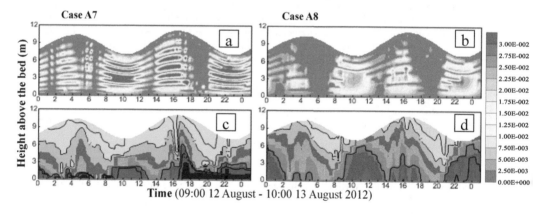

Figure 6-17. Time series of the modeled vertical eddy viscosity and SSC of Cases A7 (left panels) and A8 (right panels) at station PT2, during the spring tide.

Figure 6-18 presents modeled current velocity and SSC in the different scenarios, among which turbulence damping effects are compared. Due to the turbulence damping effect the vertical profiles of current velocity and SSC are modulated simultaneously. But according to the stratification level of SSC, Case 11 shows results more similar to measure ones than Cases 9 and 10. *Figure 6-19* compares the vertical profiles of turbulent eddy viscosity of Case 0 (no damping), Case 3 (with original turbulence damping effect) and Case 11 (with additional damping effect) at the peak flood (a) and peak ebb (b) at station PT2. Their comparison indicates that the standard k-ε turbulence model seems to underestimate the intensity of density stratification, which additional turbulence damping effect seems to be able to improve the vertical profiles of SSC effectively.

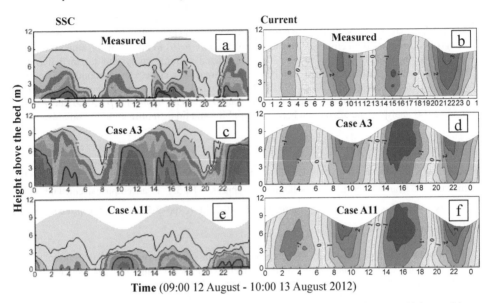

Figure 6-18. Time series of measured and modeled SSC (left panels) and current (right panels) at station PT2, during the spring tide.

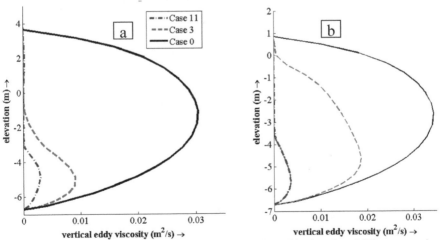

Figure 6-19. Comparisons of the vertical profiles of turbulent eddy viscosity of different scenarios at the peak flood (a) and peak ebb (b) times at station PT2.

6.3.4   Effect of drag reduction

Drag reduction means the bottom resistance (bed shear stress) of water flow is reduced due to the presence of a high concentration layer near the river bed *(Gust and Walger, 1976; Best and Leeder, 1993)*. On one hand, this high concentration layer suppresses the development of turbulence near the boundary *(Wang et al., 1998)*; on the other hand, this "soft" and "slippery" layer decreases the flow resistance ability and smoothes the bottom *(Wang, 2002)*. *Teisson et al. (1992)* investigated the influence of clear or loaded water on erosion laws in laboratory experiments and found that stratification effects were most often negligible on bottom shear stress in laboratory experiments. From the review by *Winterwerp et al. (2009)*, it is concluded that bed shear stress can be reduced in the range of 15-40% owing to the effect of drag reduction. *Wang (2002)* proposed an empirical formula (see *equation 6-10*) to allow for the smooth effect on the bed in modeling. *Wang et al. (2011)* conducted a flume study and found that SSC influences the bed shear stress significantly when SSC reaches a critical level around 0.55 kg/m³, see *Figure 6-20*. *Song and Wang (2013)* highlighted the importance of *equation 6-9* in simulation of the fine sediment transport in the Yangtze Estuary. *Winterwerp (2013)* and *Wang et al. (2014)* analyzed the interaction between high SSC and tidal amplification in estuaries.

$$C_d = \left[\frac{\kappa}{1 + AR_f \ln(z_b / z_0)} \right]^2 \qquad (6\text{-}9)$$

where, C_d is bottom drag coefficient, κ is the Von Karman constant, z_b is the layer thickness of near bed grid, , z_0 is bottom roughness, $A = 5.5$ is an empirical constant and R_f is the flux Richardson number $R_f \approx 0.25 \left[1 - \exp(-7.5 Ri_g) \right]$ *(Karimpour and Venayagamoorthy, 2014)*.

In order to evaluate the effect of drag reduction on vertical structures of the current and SSC, we implemented *equation 6-9* into the Delft3D model. Comparison of modeled SSC profile with (Case 12) and without (Case 3) the effect of drag reduction is shown in *Figure 6-21*. It can be seen that the effect reduces the overall SSC level in the water column due to reducing the drag friction coefficient, but the intensity of stratification almost does not altered, as shown in *Figure 6-22*. We should be aware that in this study, the bed deposition flux is independent of a threshold stress; if the critical deposition stress is around 0.1-0.8 Pa *(Hu et al., 2009b; Song and Wang, 2013)*, it will result in more sedimentation period and relatively higher SSC. The function of drag reduction used in modeling seems to resemble the critical erosion stress (τ_e) and erosion rate (M).

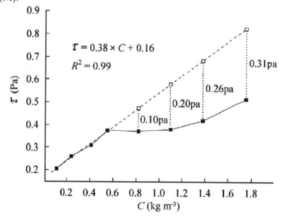

Figure 6-20.   Relationship between bed shear stress and SSC *(Wang et al., 2011)*.

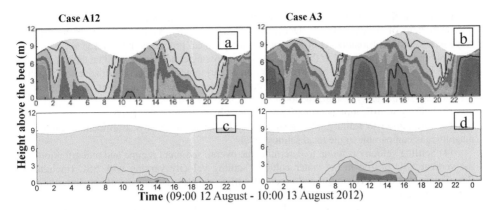

Figure 6-21. Time series of modeled SSC at station PT2 of Cases A12 (left panels) and A3 (right panels), during the spring (upper panels) and neap (lower panels) tides.

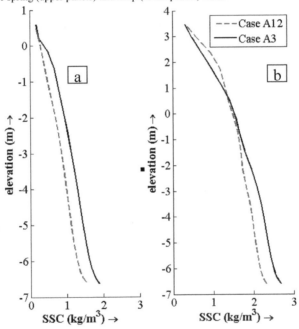

Figure 6-22. Comparisons of the vertical profiles of turbulent eddy viscosity of different scenarios at the peak ebb (a) and peak flood (b) times at station PT2.

6.4   Conclusion

The poor performance of a hydrodynamic and sediment transport model is likely to result from absence or misrepresentation of a process which critically affects the motion of current and sediment. The agreement between observations and predictions in terms of vertical profiles of current, salinity and SSC is quite challenging in a stratified flow. According to the sensitivity study which has been done, it could be found that a successful modeling of fine sediment dynamics within an ETM zone is highly depending on appropriate formulation of effects from baroclinicity, settling velocity, turbulence damping and drag reduction.

With help of a schematized 3D current-sediment-salinity model, a series of modeling responses to different treatments of the micro-scale physical processes in the ETM zone of the Yangtze Estuary are presented in this Chapter. The following conclusions could be drawn:

(1) The vertical structures of currents near an estuarine turbidity maximum zone are largely affected by saltwater intrusion, especially during lower currents, when baroclinic pressure gradient can significantly reshape the local vertical velocity profile.

(2) Salinity stratification is primarily determined by the baroclinic effects.

(3) In addition to salinity, SSC also influences the local density stratification, when its magnitude is comparable to the local salinity level.

(4) Settling velocity is found to determine the overall sediment regime and redistribution of the vertical SSC profile in the water column; settling velocity varying with SSC (including flocculation-induced acceleration and hindered settling deceleration stages) affects longitudinal sediment excursion.

(5) Vertical profiles of current, salinity and SSC within such a river plume are highly associated with the development of turbulence. The approach to modulate the vertical eddy viscosity based on the empirical dependency between gradient Richardson number (Ri_g) and Prandtl number (Pr_t) leads to a numerical instability in the stratified flow. In order to improve the SSC stratification, additional turbulence damping effect is suggested to be implemented in modeling.

Seasonal ETM variation of the Yangtze Estuary

Highlights

(1) Both tidal energy and density stratification enhance saltwater intrusion.

(2) Four independent factors determining the seasonal sediment regime are identified.

(3) River discharge impacts the horizontal and vertical distribution of residual current.

(4) Seasonally varying wind effect alters the residual currents near the riverine limit.

(5) Seasonally varying mean sea level and wind climate jointly shape the saltwater intrusion length in the DNC.

7.1 Introduction

As mentioned in *Chapter 1*, there is seasonal variability of the back-silting rate at the DNC in the Yangtze Estuary. The back-silting rate in wet seasons is 4 times than that in dry seasons. In addition, as shown in *Figures 1-5* and *1-6*, the longitudinal profiles of SSC and salinity also show an obviously seasonal variation. Along-channel stratification degree and spatial distribution of vertical SSC profiles are varied dramatically from dry season to wet season. It seems that the ETM structure is conditionally either well stratified or mixing, in different seasons. In *Chapter 3*, seasonal variation of the residual current fields was demonstrated by a 3D hydrodynamic model, in which sediment-current interaction and wind were exclusive in modeling. The numerical results showed that the residual flow regime is of critical importance for the trapping probability of sediment and the occurrence of fluid mud. In *Chapter 5*, we confirmed that the seasonal fluctuation of temperature has an effect on settling velocity of fine sediment via s series laboratory experiments. In *Chapter 6*, a sensitivity study was conducted to identify the functions of these micro-scale factors on the internal structure of flow, SSC and salinity via a schematic 3D model without macro-scale driving forces (e.g. river discharge). Modeling highlighted the importance of baroclinicity, settling velocity and additional turbulence damping on hydrodynamics and sediment transport. According to the numerical sensitivity study, we concluded that: (i) settling velocity is sensitive to suspended sediment dynamics in modeling; (ii) flocculation settlement process of fine sediments (including flocculation acceleration and hindered settling) significantly impacts the suspended sediment regime near the ETM zone; (iii) saltwater intrusion induced baroclinic effect critically influences the internal flow structure; and (iv) additional turbulence damping is an effective factor modulating SSC stratification. In short, these micro-scale physical processes are quite important for hydrodynamics and sediment transport in an estuary. Therefore, here we need to confirm the behavior of the interactions between the micro- and macro- physical processes in the Yangtze Estuary. And more importantly, the physical mechanisms governing the seasonal ETM variation are explored and analyzed.

In this chapter, firstly the seasonal differences of the circumstance in the Yangtze Estuary which may affect the current and sediment behavior are systematical distinguished. Secondly, a Delft3D-based 3D sediment-salinity-current coupled model (without morphological updating) is built up and a number of numerical experiments jointly considering the micro-scale effects and macro-scale forces are carried out. Finally, model results are discussed.

7.2 Methods

7.2.1 Investigation on seasonal variation controlling factors

For a seasonally varying estuarine system, a question arising naturally is what external and internal factors are changing or changed from season to season. For a large-scale estuary, this problem is not a simple issue. In the past, the attention associated to seasonal difference in the Yangtze Estuary has focused on the freshwater inflow. In this study, the following 5 determinants are investigated to highlight the seasonal differences in the estuary. Here, the typical representative months of dry and wet seasons are January or February and August or September in terms of the back-silting characteristics in the DNC (see *Figure 1-9* and *Table 1-3*).

(1) Freshwater inflow

Seasonal fluctuation of the Yangtze discharge is primarily due to monsoonal precipitation in the upper and middle Yangtze river basin *(Chen et al., 2001b)*. In addition, the operation of the TGD project also plays a secondary role (see *Figure 7-1*). The general operational strategy of TGD is that the reservoir starts to impound mainly after flood season (monsoon generally lasts from mid-may to September in the middle and upper reaches of the Yangtze River) and to release the stored water pre-flood for balancing the yearly water resource distribution. Normally, from May to October is the typical wet season in the middle and lower reach of the Yangtze River, mainly taking into account the run-off proportion (see *Table 7-1*). The ratio of the total run-off in the wet season to that in the dry season is 68:32, which is much smaller than the seasonal back-silting ratio (86:14) in *Table 1-3*.

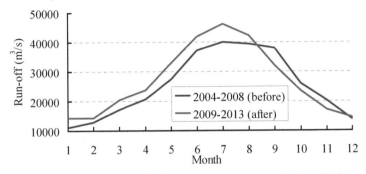

Figure 7-1.   Comparison of monthly run-off at Datong station before and after the TGD impoundment.

Table 7-1.   Monthly run-off and its proportion in one year at Datong station from 2004 to 2013. (Red data represent the wet season of the Yangtze River, while the black data is the dry season.)

Month	Jan.	Feb.	Mar.	Apr.	May	Jun.
Run-off (m³/s)	12571	13593	18750	22145	30111	39478
Proportion (%)	4	4	6	7	10	13
Month	Jul.	Aug.	Sep.	Oct.	Nov.	Dec.
Run-off (m³/s)	43028	40750	35025	24698	18676	14183
Proportion (%)	14	13	11	8	6	5

(2) Tidal level

Due to the nonlinear interactions between freshwater inflow and oceanic tides (including astronomic tides (short period), atmospheric tides (long period) and shallow water overtides) in the Yangtze Estuary, the fluctuations of tidal range and mean sea level show obvious semiannual and annual variation, respectively (*Figures 7-2* and *7-3*). The seasonal dividing point of tidal range is around June (*Figure 7-2*). The dry-wet seasonal mean sea level difference is about 30-50 cm around the nearshore region of the Yangtze Estuary (*Figures 7-3* and *7-4*).

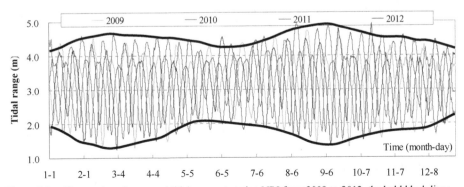

Figure 7-2.   Time-series of measured tidal range at station NPJ from 2009 to 2012, the bold black lines show overall yearly variation trend of tidal range (location of the station refers to *Figure 3-2*).

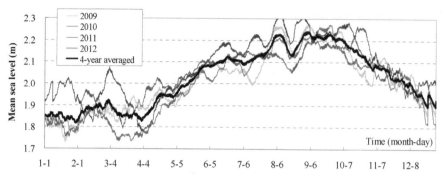

Figure 7-3. Measured mean sea level (fortnight moving averaged) at station NPJ from 2009 to 2012 (location refers to *Figure 3-2*).

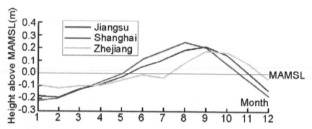

Figure 7-4. Monthly averaged sea level along the Jiangsu, Shanghai and Zhejiang coasts during the period of 1980-2013, redrawn from *www.soa.gov.cn/zwgk/hygb/zghpmgb (SOA, 2014)*, MAMSL is multi-year averaged mean-sea-level.

(3) Water temperature

For hydrodynamic and sediment transport issues, the influence of water temperature is usually neglected. But for fine sediment dynamics, temperature has a significant impact on the flocculation settling process (the details see *Section 5.3.3*). *Figure 7-5* shows monthly water temperature at station NPJ. The typical dry and wet season water temperature is about 7°C and 27°C, respectively.

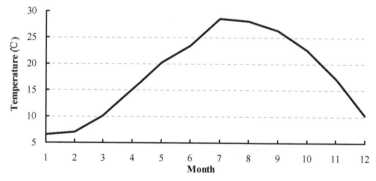

Figure 7-5. Monthly measured water temperature at station NPJ (the data are averaged from the period of 2005-2010).

(4) Wind and wave

Figure 7-6 and *Table 7-4* show 4-year series of observed wind speed and direction at NCD. Wind speed does not show obvious seasonal fluctuations. Wind speed in the winter is a little greater than that in the summer. Wind speed in January and February is about 7 m/s, and in

August is about 6 m/s. However, the prevailing wind direction from April to August is southeasterly, and from September to March is northerly.

Yearly-averaged significant water height and period at NCD are about 0.6-0.7 m and 5s. The wave climate in the Yangtze Estuary has no significant seasonal variation.

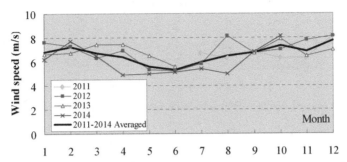

Figure 7-6. Measured monthly wind speed at station NCD from 2011 to 2014 (location refers to *Figure 3-9*).

Table 7-2. Measured monthly wind direction frequency (%) distribution at station NCD from 2011 to 2014, the numbers in red color denote the dominant wind direction of each month.

Month/Direction	Jan	Feb	Mar	Apr	May	Jun	Jul	Aug	Sep	Oct	Nov	Dec
N	18.33	19.10	14.13	8.91	6.48	3.99	3.01	6.32	14.17	16.72	15.56	18.02
NNE	5.65	8.72	7.05	4.05	3.11	2.70	1.96	4.05	8.87	6.30	4.68	4.74
NE	7.06	8.11	9.28	7.59	6.16	6.53	4.32	7.82	13.14	12.41	7.02	5.63
ENE	2.79	2.88	3.80	2.99	3.64	3.64	2.83	4.16	4.59	5.40	2.58	2.13
E	4.81	4.23	6.17	6.88	6.59	8.73	6.07	7.75	7.80	7.24	5.78	4.13
ESE	2.28	3.38	4.87	6.16	7.58	7.92	5.00	6.06	4.89	3.87	3.59	2.00
SE	4.87	7.10	11.45	18.70	21.24	20.44	16.61	17.20	10.42	6.97	7.15	4.17
SSE	2.18	3.63	6.32	11.07	12.27	13.20	21.36	13.88	6.24	4.42	4.60	2.95
S	2.89	4.38	6.19	10.19	11.20	12.65	20.11	12.62	5.53	4.19	4.41	3.38
SSW	1.18	1.21	1.85	1.81	2.30	4.13	5.02	2.98	1.20	1.18	1.35	1.35
SW	1.93	2.42	3.20	3.19	3.35	4.49	4.16	2.89	1.54	1.87	2.40	2.81
WSW	1.34	1.52	1.44	1.57	1.45	1.58	1.92	1.31	0.80	0.61	1.19	1.56
W	2.71	2.09	2.17	2.55	2.16	2.16	2.02	1.69	1.15	0.88	1.74	2.96
WNW	5.96	3.71	3.11	2.41	1.87	1.46	1.08	2.43	1.97	2.35	4.29	5.60
NW	19.85	12.84	8.72	6.27	5.73	3.18	2.04	4.56	8.63	13.12	18.50	22.92
NNW	15.09	13.71	9.29	4.63	3.99	2.47	1.45	3.44	8.42	11.38	13.96	13.78
C	1.08	0.99	0.97	1.04	0.88	0.72	1.04	0.83	0.64	1.09	1.19	1.89

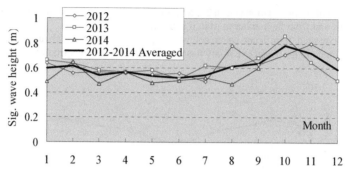

Figure 7-7. Measured monthly significant wave height at station NCD from 2012 to 2014 (location refers to *Figure 3-2*).

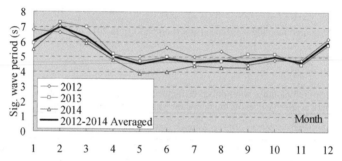

Figure 7-8. Measured monthly significant wave period at station NCD from 2012 to 2014 (location refers to *Figure 3-2*).

(5) Longshore current

Longshore current is an ocean current that moves parallel to coasts. For a larger range of areas, the macroscopic longshore current (comparatively speaking, the microscopic longshore current mainly generated in the surf zone and on tidal flat) is mainly caused by wind effect and large scale water level gradients. *Ma et al. (2010)* concluded that the seasonal variation of mean currents (see *Figure 7-9*) in the East China Sea is mainly induced by the Kuroshio Current and monsoonal forcing. In the winter, the Yangtze river plume and diluted water spread southerly along the Zhejiang coast under the northerly winter monsoon; while in the summer, the plume and diluted water turn northeasterly under the influence of southerly summer monsoon *(Chang and Isobe, 2003; Wu et al., 2011)*. Generally, the longshore current in the winter favors sediment deliver from the North Passage to the open sea and in the summer it will restrain the outflow of the Yangtze river plume to some extent (see *Figure 7-10*).

The influence of the longshore current on hydrodynamics and sediment transport in an estuary is always overlooked, because the longshore current is produced by large-scale shelf currents circulation and oceanic climate (wind, pressure, temperature, precipitation, evaporation, etc.). However, for a large scale estuary, the Yangtze Estuary, the relationship between seasonal longshore current and sediment delivering from river to sea should be studied to enrich our knowledge on sediment regime in this region.

In summary, the typical dry-wet seasonal differences are summarized in *Table 7-3*. Seasonal difference of the longshore current is simply represented via the seasonal variation of wind and mean sea level jointly.

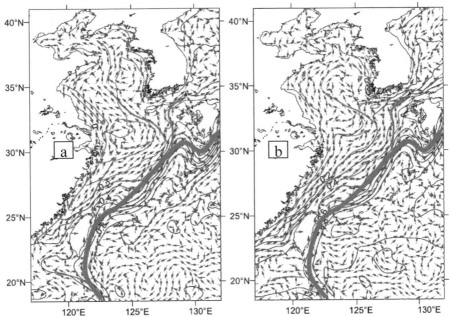

Figure 7-9. Dry (a) and wet (b) seasonal longshore and residual current field in the East China and Yellow Sea, the bold arrows denote the dominant directions (source: Figure 3 of *(Ma et al., 2010)*).

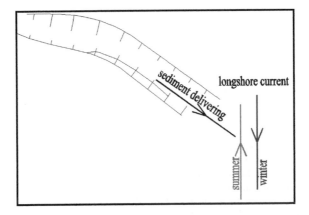

Figure 7-10. Schematic relationship between longshore current and sediment delivering.

Table 7-3. Identified typical seasonal controlling factors of the Yangtze Estuary.

Factors	Dry season	Wet season
Freshwater inflow (m³/s)	13,000	43,000
Wind speed (m/s) /direction (°)	7/0	6/145
Mean sea level (m)	1.8	2.2
Water temperature (°C)	7	27

7.2.2 Field observations

Two field campaigns for comparing the seasonal variations of ETM in the North Passage were deployed in the dry and wet seasons of 2012. The current, salinity and SSC observations in ten ship-borne stations (*Figure 7-11*) along the DNC were carried out during both spring and neap tides. These data are collected hourly at six layers in water column with relative depths of 0.05(near surface), 0.2, 0.4, 0.6, 0.8, and 0.95 (near-bed) at each station. The observation time of dry season for neap tide is from 07:00, 17-2-2012 to 08:00, 18-2-2012 and for spring tide is from 06:00, 23-2-2012 to 07:00, 24-2-2012. And the observation time of wet season for neap tide is from 09:00, on 12-8-2012 to 10:00, 13-08-2012 and for spring tide is from 06:00, 17-8-2012 to 07:00, 18-8-2012.

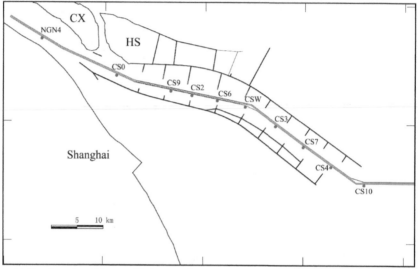

Figure 7-11. Location of the ten ship-born stations.

Figure 7-12 presents the measured tidal-averaged SSC and salinity along the DNC under different tidal and seasonal conditions. Firstly, regarding the seasonal variations, one can see the following features. (i) The overall SSC in the wet season is higher, more horizontally centralized and vertically stratified than that in the dry season. (ii) Meanwhile, it should be noted that the contour lines of salinity in the wet season are denser than those in the dry season. It means the along-channel gradient of salinity in the wet season are much larger compared to that in the dry season. The longitudinal density gradient will introduce baroclinic pressure gradient forcing (see *equation 2-6*). Therefore, it further indicates that the baroclinicity effect (up-estuary trend of mass transport) in the wet season is stronger than that in the dry season, especially from the region CS9 to CS4 (including the highlighted reach in *Figure 1-10*). On the other hand, in comparison with the slope of the salinity contour lines in dry and wet seasons, the smaller slope denotes higher vertical density gradient, and larger vertical density gradient means sharper density stratification and lower mixing efficiency. Therefore, the vertical mixing status of fluid in the wet season is not beneficial for the dispersion of near-bed high SSC layer. (iii) Generally, the saltwater intrudes more up-estuary in the dry season, due to lower freshwater inflow.

Secondly, considering neap-spring variations, the following characteristics are extracted. (i) SSC during the spring tide is obviously higher that during the neap tide. It indirectly means the SPM in the water column is mainly controlled by the current energy. Thus, tidal-induced sediment resuspension, entrainment and transport are the primary mode of sediment motion in the North Passage. (ii) The slope of isohalines during the neap tide is steeper than that during the spring tide, which means the density stratification during the neap tide is sharper. Therefore the baroclinic force (up-estuary direction) is larger and the turbulence mixing is weaker during the neap tide. So the sediment trapping efficiency in neap tide is greater comparing to that in spring tide. From above two aspects, it can be inferred that the sedimentation materials in the DNC are mainly eroded during spring tides. And they will be distributed by the flow regime and settle down during neap tides. (iii) Referring to the intra-tidal comparison, because of higher tidal energy, saltwater intrudes more landward during spring tide normally *(Mitchell et al., 2006)*. However in the wet season, the 4 psu isohalines are located more landward near the bottom during the neap tide and more seaward near the surface during the spring tide. That means both tidal energy and density stratification favors landward saltwater intrusion and sediment transport.

The phenomenon describing the seasonal ETM variation is quite clear in *Figure 7-12*, however the question "which factors (micro- and macro scale) governing the seasonal variation of ETM in the North Passage of the Yangtze Estuary" is still unsolved. Therefore, we study the underlying physical mechanisms via numerical modeling.

Multiscale physical processes of fine sediment in an estuary

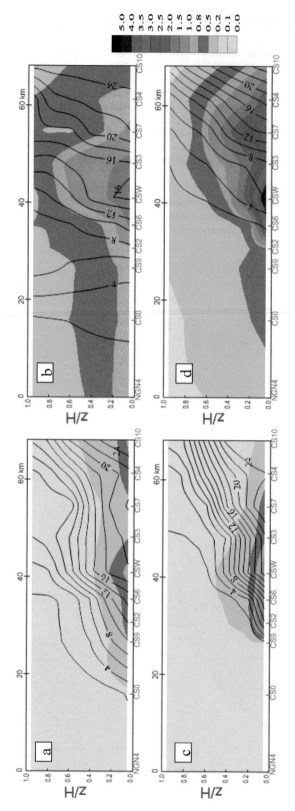

Figure 7-12. Measured tidal-averaged SSC along the DNC during the neap (a c) and spring (b d) tides of dry (a b) and wet (c d) season, unit kg/m³. Isohalines are given in black lines, the interval is 2 psu.

7.2.3   Numerical modeling

7.2.3.1   Model descriptions

We examine the seasonal variations of fine suspended sediment dynamics in the Yangtze Estuary through the Delft3D model *(Deltares, 2014)*. Owing to its open-source accessibility, it is convenient to adjust and modify some physical processes which have been formulated and parameterized in the standard model (version 6.01), especially associated to sediment-current interactions, which is quite case-specific and highly relies on various methodologies of understanding and characterizing mud dynamics.

The model domain *(Figure 7-13)* extends from the Datong (freshwater inflow boundary) to the coastal area with the borders (170 km on the north side, 300 km on the east side and 120 km on the south side). Bathymetry is shown in *Figure 7-14*. The N-shaped channel in the upstream is designed to simplify the complex land boundary (see *Figure 3-13*) and bathymetry. Comparing to specifying tidal level at the upstream, this treatment is recommended in a riverine dominated estuary to ensure the accuracy of total volume of the propagated tidal prism. The topography and bed resistance of the N-shaped channel should be adjusted according to the validation of nearby water level and current.

To account for mimicking the dredging trench (DNC, the channel width is 250 m) in the Yangtze Estuary model, the horizontally orthogonal curvilinear mesh (410×601) is locally refined in the North Passage, see *Figure 7-13*. The minimum size of the mesh is about 80 m. The vertical grid is divided into 10 sigma layers (the relative thicknesses are 7%, 10%, 13%, 16%, 16%, 12%, 10%, 8%, 5% and 3% from surface to bottom, respectively). In the model, the jetties and groins are represented by the local weir method *(Deltares, 2014)*. The open boundary is forced by 16 astronomically tidal components (M2, S2, N2, K2, K1, O1, P1, Q1, MU2, NU2, T2, L2, 2N2, J1, M1, and OO1) derived from the NaoTide data set *(Matsumoto et al., 2000)* (*www.miz.nao.ac.jp*). For the river boundary, wind and mean sea level data see *Table 7-3*.

The micro-scale effects of baroclinicity, settling velocity and additional turbulence damping, which have confirmed in *Chapter 6*, are all included in the model. In the estuarine system, suspended sediment is the dominant transport mode and SPM is mainly obtained by regional erosion and resuspension *(van Rijn, 2007)*. Only the area encircled by red lines is erodible, and the critical erosion shear stress outside the area is set to 1000 N/m^2.

The model is warmed up from a cold start condition for 90 days in both the wet and dry seasons mainly in order to get an initial condition for salinity. To prevent the effect of SSC on fluid density making noise in hydrodynamics and the initial horizontal salinity distribution, the erosion rate in the spin-up cases is set to be an order of magnitude lower. For calibration and numerical experiment cases, the simulations are run for 30 days from an initial condition. The physical parameters are calibrated (*Table 7-4*).

7.2.3.2   Scenario configuration

In this section, two runs for storing initial condition, two validation cases and eight numerical experiments (see *Table 7-4*) were conducted firstly to confirm the effects of settling velocity and additional turbulence damping on fine sediment dynamics in the complex estuarine system (Cases B1-B4). Subsequently we investigate the influences of the macro-scale factors (freshwater inflow, mean sea level and wind) on sediment regime (Cases B5-B8). Especially, we focus on their effects on the vertical internal structures of flow and SSC within the river plume in typical dry and wet seasons. Through analyzing the characteristics of seasonally varying longitudinal ETM excursion and residual current, the functions of these multiscale physical processes are investigated.

Table 7-4. List of configurations of the numerical experiments.

Case No.	Description	Common setting
I-dry	Spin-up for 90 days with typical freshwater inflow and typical wind climate of dry season (see *Table 7-2*) with a lower erosion rate (1e-6)	
I-wet	Spin-up for wet season (see *Table 7-2*) with a lower erosion rate (1e-6)	
Calibration-dry	With actual freshwater discharge and wind climate in Feb. of 2012.	Time step: 1 min
Calibration-wet	With actual freshwater discharge and wind climate in Aug. of 2012.	Manning coefficient: 0.01-0.02, varied with
B1	Constant fall velocity 0.4 mm/s	bathymetry
B2	With the experimental Yangtze mud fall velocity (for dry season), standard k-ε model	Horizontal viscosity and diffusivity: 0.1 Background vertical viscosity
B3	With the experimental Yangtze mud fall velocity (for wet season), standard k-ε model	and diffusivity: 1e-5 Critical erosion shear stress: 0.1-0.3 N/m² varied
B4	scaled k-ε model, only viscosity is amplified based on Ri (vertical averaged Ri)	from XLJ to NPJ Critical deposition stress: 1000 N/m²
B5	Wet season discharge + dry season wind climate	Erosion rate: 3e-5
B6	Dry season discharge + wet season wind climate	
B7	Wet season discharge and MSL + dry season wind	
B8	Dry season discharge and MSL + wet season wind	

Figure 7-13. Google Earth based model domain and horizontal mesh (including local enlarged grid of the North Passage), upstream riverine watercourse is simplified to an N-shaped channel for preserving the conservation of tidal prism.

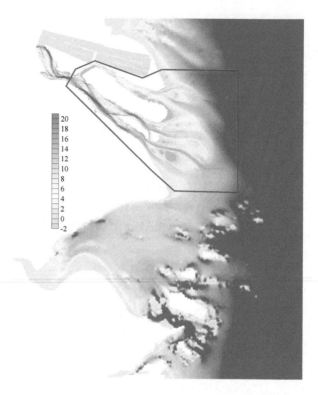

Figure 7-14. Bathymetry of the Yangtze Estuary, the area enclosed by red lines is the domain of the erodible riverbed.

7.2.3.3 Validation

Validation is an essential procedure to confirm whether a model fits the measured data or analytical solution well or not, and to verify that the simulation can catch the main specified characteristics of a targeted system. The measured data (water level, current (6-layers), SSC and salinity) from neap tide to spring tide in the wet season are calibrated against modeled results. *Figures 7-15* to *7-18* present comparison between observed and modeled results. The performance of the validation indicates that the model showed a reasonable capability.

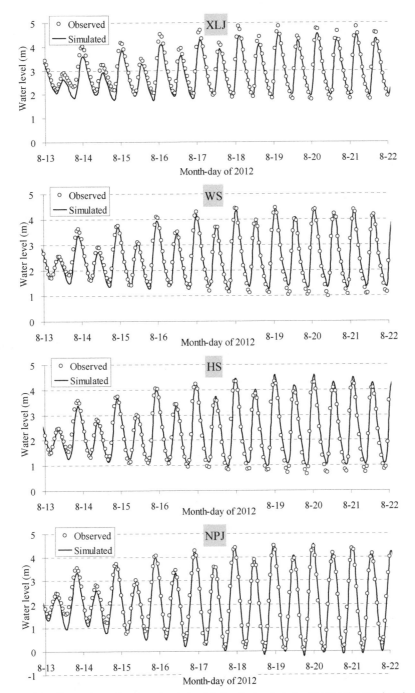

Figure 7-15. Comparison of measured and simulated water levels in the Yangtze Estuary, location of the tidal gauging stations refers *Figure 2-3*. Time intervals of measured and modeled results are 1 h and 10 min, respectively.

Multiscale physical processes of fine sediment in an estuary

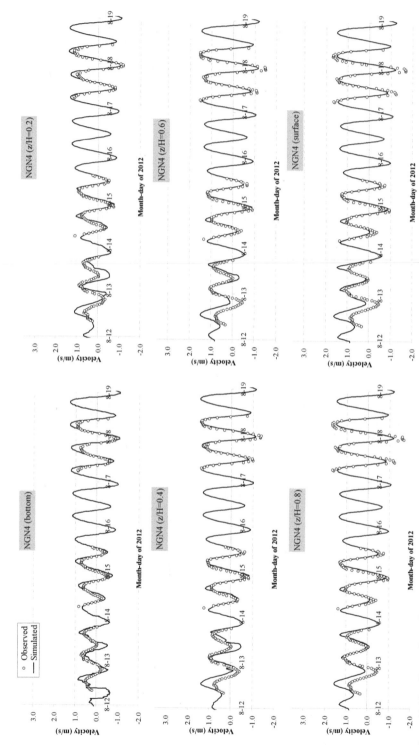

Figure 7-16. Comparison of measured and simulated current velocity at different relative depths in NGN4, CSW and CS10 (the three stations are located upper-, middle- and lower reaches of the North Passage), location of the stations refers *Figure 7-11*. Time intervals of measured and modeled results are 1 h and 10 min, respectively.

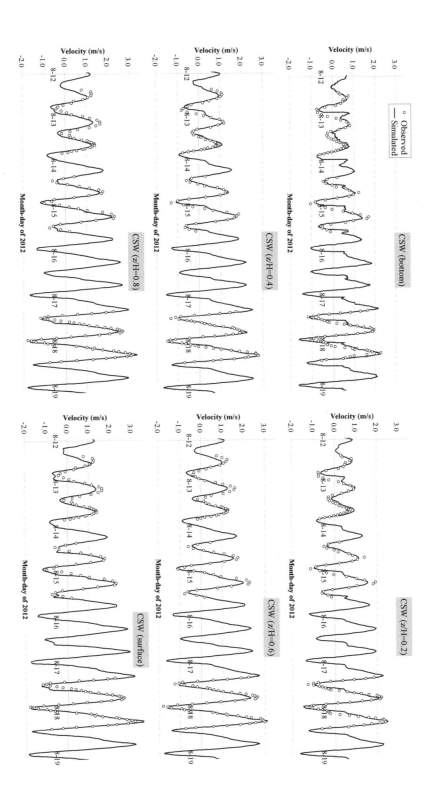

Figure 7-16. Continued

Multiscale physical processes of fine sediment in an estuary

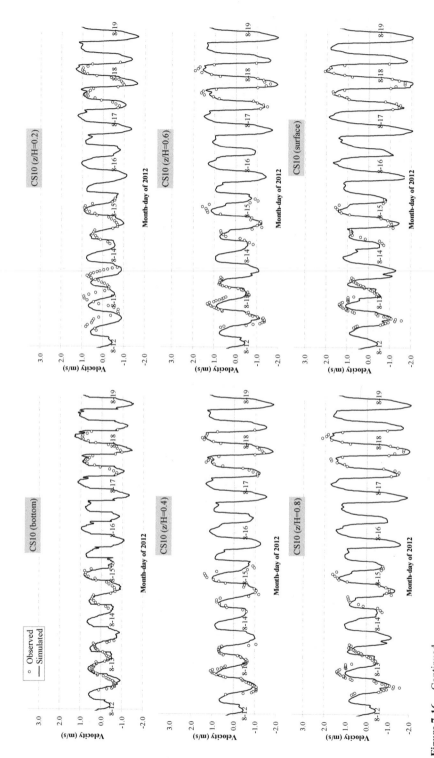

Figure 7-16. Continued

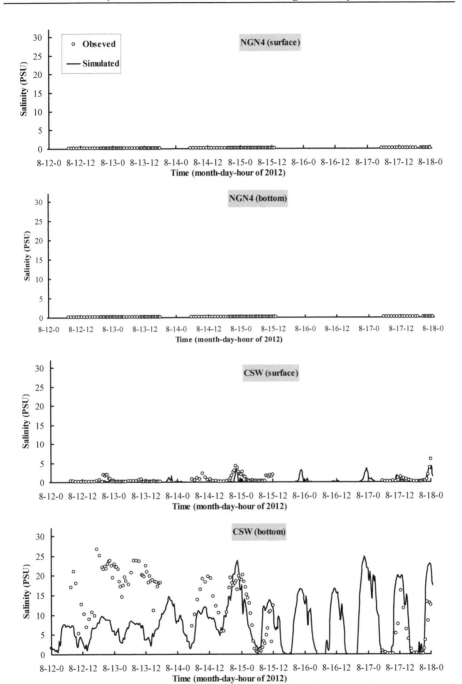

Figure 7-17. Comparison of measured and simulated salinity near surface and bottom layers at the NGN4, CSW and CS10 (the three stations are located upper-, middle- and lower reaches of the North Passage), location of the stations refers *Figure 7-11*. Time intervals of measured and modeled results are 1 h and 10 min, respectively.

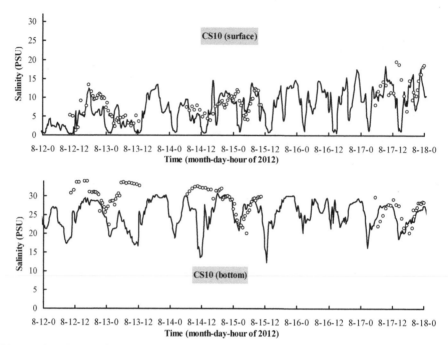

Figure 7-17. Continued

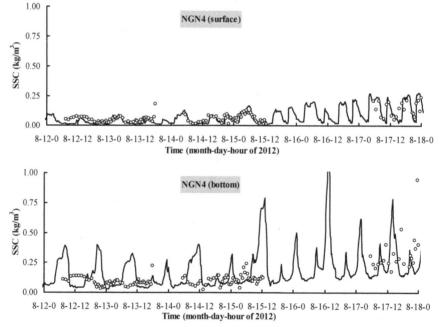

Figure 7-18. Comparison of measured and simulated SSC near surface and bottom layers at the NGN4, CSW and CS10.

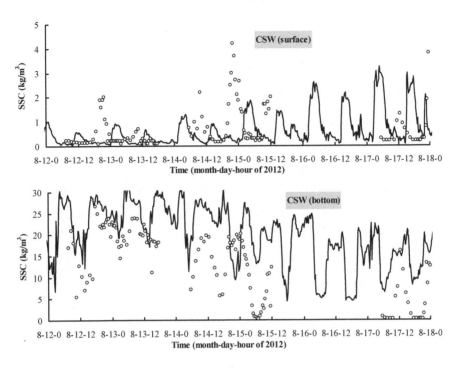

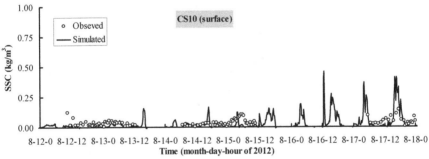

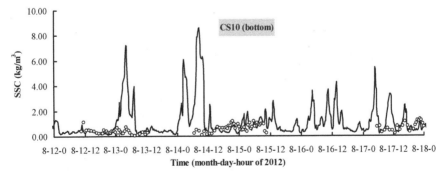

Figure 7-18. Continued

7.3 Results and discussions

Consider the four independent parameters (see *Table 7-3*) as the driving force of the seasonal ETM variation, the effects of river discharge, wind speed and direction, Seasonally varying settling velocity and mean sea level on hydrodynamics, salinity and fine sediment dynamics are examined via the validated Yangtze Estuary model.

7.3.1 River discharge

The runoff at the Datong (location see *Figure 1-1*) is commonly considered as the primary controlling factor of the wet-dry seasonal marine environmental variation in the Yangtze Estuary *(Chen et al., 2006; Hu et al., 2009b)*. Variation of river discharge directly alters the nonlinear competition between seaward river flow and landward tidal propagation in an estuary. Residual current is a favorable indicator to present the seasonal difference. In *Section 3.3.2*, the distinct horizontal characteristics of tide-induced horizontal residual current fields in the typical wet and dry seasons were presented and analyzed; as shown in *Figure 7-20a*. So, here we mainly focus on the vertical residual currents profile in the DNC (*Figure 7-19*).

According to *Figure 7-19*, we can see evidently that, (i) even the discharge of riverine inflow increases obviously from 13,000 m^3/s in the dry season to 43,000 m^3/s in the wet season, the seaward residual intensity near the river bed is decreased from ebb-dominant to flood dominant, especially near the salt front (locations see *Figure 7-11*, from CS6 to CS4) and (ii) the residual current is relative weaker in the neap tide than that in the spring tide. That means the residual currents near the riverbed are not beneficial the sediment delivery in the DNC. Those fine sediments tend to deposit and accumulate in the convergent area, from CS6 to CS4 in the wet season. Although the tidal energy is relatively weak in the dry season, the residual current is continuously ebb-dominant, so the sediment could be transported to the sea easily.

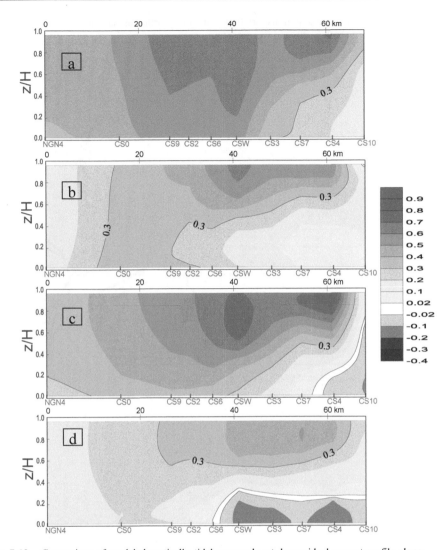

Figure 7-19. Comparison of modeled vertically tidal-averaged wet-dry residual current profile along the DNC, red color means ebb-dominant and blue means flood-dominant, unit: m/s. (a) is the residuals during the neap tide of dry season, (b) is in the spring tide of dry season, and (c) is the residuals in the neap tide of wet season, and (d) is in the spring tide of wet season.

7.3.2   Wind effect

Li et al (2012) found that the saltwater intrusion of the Yangtze Estuary is significantly influenced by wind speed and direction. According to the modeling results of this study, the following three functions of wind effect are found.

(1) Wind effect significantly changes the residual pattern outside the DNC (*Figure 7-20*), where is the riverine limit of the North Passage (it means the hydrodynamics is less influenced by river forcing and the current is just changed from reciprocating-dominated (to-and-fro flow) to rotational flow). Under the same river discharge and tidal forcing (*Figures 7-20b, c*), the wet seasonal southeasterly wind in both the wet and dry seasons favors forming a clockwise residual vortex, which is not beneficial for riverine sediment delivery. While the dry seasonal north wind

145

is favorable for sediment transportation from the North Passage and South Passage to the Zhejiang coasts both in dry and wet seasons.

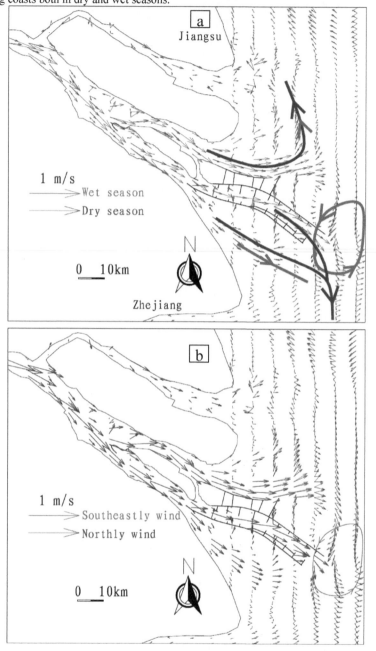

Figure 7-20. Vertical- and fortnight-averaged residual currents distribution in the Yangtze Estuary. Bold curves denote the dominant transport direction of water mass. (a) is the residuals of typical wet (with southeasterly wind and larger runoff 43,000m³/s) and dry (with northerly wind and lower runoff 13,000m³/s) seasons, (b) is the residual flow field of typical wet season with different wind climates (red arrows denote with southeasterly wind and blues denote with northerly wind), and (c) is the residual flow field of typical dry season with different wind climates.

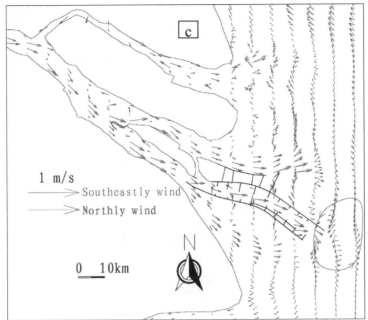

Figure 7-20.   Continued.

(2) Wind effect impacts up-estuary saltwater intrusion of the Yangtze Estuary. According to the result of the numerical experiments (*Figure 7-21*), comparing to the wet seasonal southeasterly wind (*Figures 7-21a, d*), the dry seasonal wind (north wind) climate (*Figures 7-21b, c*) is in favor of up-estuary saltwater intrusion.

(3) Wet seasonal southeasterly wind in the summer enhances the saltwater intrusion and stratification (see *Figure 7-22*). With the typical summer southeasterly wind effect included in the Yangtze estuarine system, the intensity of saltwater intrusion is much greater than that without wind effect. And the effect is stronger during the neap tide, when the currents are relatively weak and the southeasterly wind drag force will enhance the up-channel excursion of saline water.

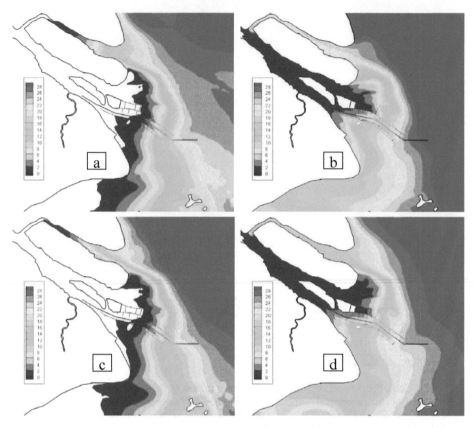

Figure 7-21. Comparison of modeled vertically tidal-averaged salinity (the colored contours are cutoff if the value below 0.1, unit: psu). (a) is the salinity distribution in the dry season with southeasterly wind, (b) is in the dry season with northerly wind, (c) is in the wet season with northerly wind, and (d) is in the dry season with southeasterly wind.

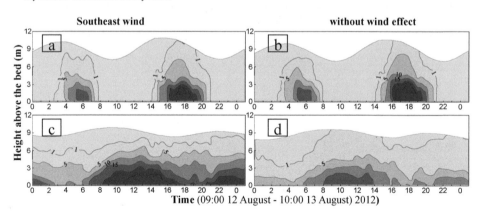

Figure 7-22. Comparison of simulated salinity at CS3 (location refers *Figure 7-10*) with (left panel) and without (right panel) wind effect during the spring (upper panel) and neap (lower panel) tides in the wet season.

7.3.3   Effect of mean sea level

Effect of mean sea level is tested by the Case B5. In Case B5, mean sea levels at all the sea boundaries (amplitude of A0 tide component) are set to 1.8 m, but the river discharge is matched with the typical runoff in the wet season. The primary influence of the mean sea level on the system is that it impacts the saltwater intrusion length. From *Figures 7-23*, in the wet season, if the mean sea level at the boundaries is decreased from 2.2 m to 1.8 m, the saltwater intrusion length will be extended landward about 5-15 km. The seasonal fluctuation of mean sea level is directly induced by the long period (semi-annual) atmospheric tides and the shelf currents in the East China Sea *(Wu et al., 2011)*. Mean sea level controls the saltwater intrusion length in the North Passage.

Saline water intrusion induced baroclinic effect and density flow in the estuary will reshape internal flow structure and the vertical SSC profile. The diluted water *(Wu et al., 2010; Wu et al., 2011)* circulation in the Yangtze Estuary not only impacts the availability of freshwater resource, but also locates the position of salt front and turbidity plume. As for a further consequence, it affects short-term and long-term river regimes and estuarine morphological evolutions.

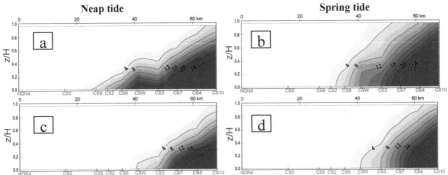

Figure 7-23. Comparison of along-channel salinity profiles under various mean sea level in the wet season. The left panel is the longitudinal salinity profile of the neap tide, and the right panel is the profile of the spring tide; the mean sea level in (a) and (b) is 2.2 m, and in (c) and (d) is 1.8 m.

7.4   Conclusion

In this chapter, firstly, a comprehensive investigation on the seasonally-varying factors in the Yangtze Estuary is carried out. Four independent determinants, including discharge of riverine inflow, wind speed and direction, mean sea level and water temperature are identified as the active factors for the seasonal ETM variation in the Yangtze Estuary.

Secondly, measured data describing the characteristics of neap-spring and seasonal variations of salinity and SSC are presented. It shows that the vertical mixing status of fluid in the wet season is not beneficial for dispersion of the near-bed high SSC layer, especially during the neap tide. Both higher tidal energy and sharper density stratification favors saltwater intrusion more landward.

Finally, the Delft3D based Yangtze Estuary model was built to investigate the interaction between micro-scale and macro-scale physical processes of fine sediment and their influence on seasonal ETM dynamics. According to the numerical sensitivity study, we found that: (1) River discharge impacts both the horizontal and vertical residual pattern of an estuary. (2) The wet seasonal southeast wind favors forming a clockwise residual vortex, which is not beneficial for riverine sediment delivery, while the dry seasonal north wind is favorable for sediment transportation from the North Passage and South Passage to the Zhejiang coasts. (3) Seasonally varying mean sea level shapes the saltwater intrusion length in the DNC.

Conclusions and outlook

Highlights

(1) Multiscale physical processes jointly affect current and sediment regime in a fine sediment estuarine system.

(2) Multi-approaches (field survey, laboratory experiment and numerical modeling) benefit characterizing a complex physical phenomenon or process.

(3) Multi-angles from data-driven analysis and process-based modeling will advance our understanding comprehensively.

8.1 Conclusions

The study aiming to improve the understanding of fine sediment physical processes in a complex estuary is conducted via various approaches, including field survey (*Chapter 2, 3, 4, 6 and 7*), laboratory experiments (*Chapter 5*) and numerical modeling (*Chapters 2, 3, 6 and 7*). In the following a number of aspects are discussed to achieve an enhanced understanding of the serious back-silting problem (see *Section 1.2.2*) in the DNC of the Yangtze Estuary.

(1) During the channel deepening and narrowing processes (induced by engineering and dredging works from 1998 to 2010), the North Passage experienced a regime shift towards sharp stratification and high sediment trapping efficiency conditions. In *Chapter 2*, measured and modeled historical evolutions in hydrodynamics, salinity and SSC are evaluated and linked to the channel deepening processes. From the analysis, on the one hand, the estuarine environmental gradients (the salinity and SSC) have been compressed, and the fresh-salt gradient became steeper. This has had an indirect effect (strengthening the stratification effect near the ETM area and enhancing the up-estuary sediment transport) on the waterway siltation. On the other hand, the traditional strategy to increase navigable water depth and minimize waterway sedimentation around the outlet (mouth) of an estuary is to increase current intensity via engineering construction or dredging, which is inherited from the way of river regulation. The shortcoming of this strategy is that it will induce landward net sediment transport near the riverbed and run a risk of high turbidity. The related understanding is also getting more and more recognized recently by other scholars *(Winterwerp, 2011a; Winterwerp, 2013; Wang et al., 2014)*.

(2) Fluid mud is a special phenomenon of the motion of suspended sediment, and it is also a typical example to attain a better knowledge of the physical processes of fine sediment than normal suspended sediment dynamics. Based on measured and modeled results, with regard to the mechanisms of storm-induced fluid mud, we realized that (i) enough SPM is available by wave agitation from adjacent shoals, and they have to oscillate within the upper reach of the channel controlled by local hydrodynamics for a considerable period; and (ii) downslope transport of fluid mud from the flanks to the deepened navigational channel possibly reduces diffusion of the fluid mud, and provides large quantities of sediment sources for fluid mud; this effect may be enhanced with the channel deepening. (iii) the fluid mud dynamics process is an advective phenomenon determining the central position of fluid mud layer along the channel, and also it is a tidal energy-influenced phenomenon, controlling the dissipation or growth of fluid mud; and (iv) both suspended particulate matter availability and local residual flow regime are of critical importance for the trapping probability of sediment and the occurrence of fluid mud.

(3) Observed data presenting the spatial and temporal spring–neap variations of velocity, salinity and SSC in the DNC of the Yangtze Estuary during the 2012 wet season (*Chapter 4*) show the distinct characteristics of fine sediment dynamics. They are (i) Field observations indicate that the transport of residuals generated by internal tidal asymmetry plays a dominant role in maintaining a stable density stratification interface near to the ETM. (ii) During a single tidal cycle, the longitudinal scale of the channel is sufficiently long so that the ETM cannot be displaced outside of its geographic boundary, and the convergent zone of residuals during both the neap and spring tides oscillates in the middle and lower reaches of the DNC. This encourages the formation of a near-bed high-SSC layer, which favors siltation in the dredged channel. (iii) Stratification and turbulence damping effects near to the ETM induce the upper half (surface layer) of the estuary to be ebb-dominant and the lower part (near bed) to be flood-dominant, which is a previously undocumented phenomenon in this region. (iv) The residual pattern of currents, salt flux and sediment flux are non-similar in the stratified estuary, and the salinity-induced baroclinic pressure gradient is one of the dominant factors that control variation of vertical velocity structure. Additionally, residual analysis suggests that tidal resuspension and lateral sediment supply contribute to providing a sediment source during spring tides in the wet seasons.

(4) Quantitative and qualitative characterization of the flocculation settling processes of fine

sediments is of critical importance for understanding fine sediment dynamics in an estuary. The parameter of settling velocity (SV) in numerical modeling is quite sensitive to sediment regime. Our experimental results showed that SSC, salinity and temperature all affect SV, but to different extents. The relationships between SV of estuarine fine sediments and its various determinants (e.g., SSC, salinity and temperature) are highly dependent on specific environmental conditions. Furthermore, each determinant has its own impact weights in various flocculation stages. During the first stage (accelerating flocculation settling), SSC is a dominant factor; conversely, in the second stage (maximum flocculation settling), dominant weights are hold by salinity and temperature. And more important, in the third stage (hindered settling), SV is least sensitive to all of these determinants compared to the first two stages.

(5) In terms of the micro-scale physical processes within the river plume of an estuary, the modeled data from a small-scale schematic 3D model demonstrated that: (i) Saltwater intrusion mainly controls the vertical structure of currents near an estuarine turbidity maximum area, especially during the times of lower current speed, when the baroclinic pressure gradient forces can significantly reshape local vertical velocity profile. (ii) With regard to salinity, SPM helps controlling local density stratification, when the magnitude of SSC is comparable to the local salinity level, and thus the SPM-varied fluid density effect also influences vertical sediment dispersion. (iii) The main function of the settling velocity in modeling is to determine the overall sediment regime and to redistribute the vertical SSC profile in the water column; SSC-varied fall velocity (including flocculation-induced accelerated and hindered settling stages) affects vertical sediment excursion. (iv) Vertical profiles of current, salinity and SSC within a river plume are strongly dependent on a correct prediction of the development of turbulence. The approach of modifying vertical eddy viscosity based on Ri_g-Pr_t empirical dependency leads to numerical instability in stratified flow. In order to improve the SSC stratification, additional turbulence damping effect is recommended to allow for in modeling.

(6) In terms of the macro-scale physical processes in the tide- and river- dominated estuary, the numerical modeled data from a large-scale real-world 3D model evidenced that: (i) River discharge largely impacts the pattern of residual currents. (ii) Distinct seasonal wind effect alters the longshore currents. (iii) Seasonally varying mean sea level shapes the saltwater intrusion length in the DNC.

8.2 Outlook

The presented works of multiscale physical processes of fine sediment in an estuary give rise to further discussion and research. It is important to keep in mind that a few aspects in association with fine sediment transport were not addressed in this study, and a number of problems stimulated by the present findings need to be explored further in the future.

(1) No bathymetry updating is included in modeling, which means that the feedback between back-silting or morphdynamics and suspended sediment is not covered. This process relates to erosion, deposition, sediment sorting, self-weight consolidation, liquefication and variation of erodibility of river-bed.

(2) In this study, we first utilize an unstructured grid 3D model (*Chapter 3*), then we use an orthogonal curvilinear mesh 3D model (Delft3D, *Chapters 6* and *7*). The main reason for the change is due to the conservation of the advection term in the momentum equations. For a big domain and long-time simulation of estuarine circulation, the Eulerian-Lagrangian back-tracking method will impact the mass transport of the system. In the future, how to balance the computational efficiency and the conservation of the advection term is a critical numerical issue for 3D shallow water models.

(3) Since multiscale physical process of sediment are simultaneously taken into account in modeling, the sensitivity and response of the resolution of horizontal and vertical mesh is not investigated. The temporal and spatial variations of different scale of physical processes are of wide variety and different precision, a small noise in one parameter might induce large error in another process.

(4) This study has highlighted the importance of turbulence modeling. A reasonable prediction of the development of turbulence is a significant prerequisite to simulate hydrodynamics and sediment transport in high turbid and stratified flow. More efforts investigating the relationships among turbulence, density stratification and internal flow structure are expected for future modeling.

(5) Sediment research is highly empirical- and experimental-oriented, some key parameters are highly site- and case-specific. Therefore, empirical formulas that rely on environmental factors are limited in their applicability to different field conditions.

(6) The interaction between flocculation and stratification needs more efforts to improve and transfer the poor theoretical understanding of complex interactions between currents, sediment and other forces near the bed in the channel.

(7) The relation between the active engineering-induced tidal amplification (the engineering works is aiming to increase the seaward sediment deliver ability and minimize waterway siltation) and the passive tidal amplification as a result of the regime shift towards hyper-turbid condition and density stratification (in this case, it will result in landward sediment transport near the bottom) in an estuary need to be distinguished. And the methodology to regulate a riverine and estuarine waterway should be identified and be dealt with different strategies.

References

Abarbanel, H.D.I., Holm, D.D., Marsden, J.E., Ratiu, T., 1984. Richardson Number Criterion for the Nonlinear Stability of Three-Dimensional Stratified Flow. *Physical Review Letters*, 52(26), 2352-2355.

Abril, G., Nogueira, M., Etcheber, H., Cabeçadas, G., Lemaire, E., Brogueira, M.J., 2002. Behaviour of Organic Carbon in Nine Contrasting European Estuaries. *Estuarine, Coastal and Shelf Science*, 54(2), 241-262.

Agrawal, Y.C., Pottsmith, H.C., 2000. Instruments for particle size and settling velocity observations in sediment transport. *Marine Geology*, 168(1-4), 89-114.

Al Ani, S., Dyer, K.R., Huntley, D.A., 1991. Measurement of the influence of salinity on floc density and strength. *Geo-Marine Letters*, 11(3-4), 154-158.

Ali, K.H.M., Geoprgiadis, K., 1991. Laminar motion of fluid mud, *Proceedings of the Institution of Civil Engineers*, 795-821.

Ali, K.H.M., Crapper, M., O'Connor, A., 1997. Fluid mud transport, *Proceedings of the Institution of Civil Engineers*, 64-78.

Bachmann, R.W., Hoyer, M.V., Vinzon, S.B., Canfield, D.E., 2005. The Origin of the Fluid Mud Layer in Lake Apopka, Florida. *Limnology and Oceanography*, 50(2), 629-635.

Bardina, J.E., Huang, P.G., Coakley, T.J., 1997. *Turbulence modeling validation, testing, and development*. NASA Technical Memorandum (110446). National Aeronautics and Space Administration, Ames Research Center.

Battjes, J.A., Stive, M.J.F., 1985. Calibration and Verification of a Dissipation Model for Random Breaking Waves. *Journal of Geophysical Research*, 90(C5), 9159-9167.

Berhane, I., Sternberg, R.W., Kineke, G.C., Milligan, T.G., Kranck, K., 1997. The variability of suspended aggregates on the Amazon Continental Shelf. *Continental Shelf Research*, 17(3), 267-285.

Berlamont, J., Ockenden, M., Toorman, E., Winterwerp, J., 1993. The characterisation of cohesive sediment properties. *Coastal Engineering*, 21(1-3), 105-128.

Best, J.L., Leeder, M.R., 1993. Drag reduction in turbulent muddy seawater flows and some sedimentary consequences. *Sedimentology*, 40, 1129-1137.

Booij, N., Ris, R.C., Holthuijsen, L.H., 1999. A third-generation wave model for coastal regions I, Model description and validation. *Journal of Geophysical Research*, 104(C4), 7649-7666.

Boonkasame, A., Milewski, P., 2012. The Stability of Large-Amplitude Shallow Interfacial Non-Boussinesq Flows. *Studies in Applied Mathematics*, 128(1), 40-58.

Brors, B., Eidsvik, K.J., 1992. Dynamic Reynolds stress modeling of turbidity currents. *Journal of Geophysical Research: Oceans*, 97(C6), 9645-9652.

Bruens, A., 2003. *Entrainment mud suspensions*. PhD thesis, Delft University of Technology, Delft, the Netherlands, 137pp.

Burchard, H., Baumert, H., 1998. The Formation of Estuarine Turbidity Maxima Due to Density Effects in the Salt Wedge. A Hydrodynamic Process Study. *Journal of Physical Oceanography*, 28(2), 309-321.

Burchard, H., Hetland, R.D., 2010. Quantifying the Contributions of Tidal Straining and Gravitational Circulation to Residual Circulation in Periodically Stratified Tidal Estuaries. *Journal of Physical Oceanography*, 40(6), 1243-1262.

Burt, T.N., 1986. Field Settling Velocities of Estuary Muds. In: A.J. Mehta (Ed.), *Estuarine Cohesive Sediment Dynamics*. Springer-Verlag, 126-150.

Buschman, F.A., Hoitink, A.J.F., van der Vegt, M., Hoekstra, P., 2009. *Subtidal water level variation controlled by river flow and tides*. Water Resource Research, 45, W10420.

Camp, T.R., 1936. A Study of the Rational Design of Settling Tanks. *Water Environment Federation*, 8(5), 742-758.

Canuto, V.M., Howard, A., Cheng, Y., Dubovikov, M.S., 2001. Ocean Turbulence. Part I: One-Point Closure Model-Momentum and Heat Vertical Diffusivities. *Journal of Physical Oceanography*, 31(6), 1413-1426.

Casulli, V., Walters, R.A., 2000. An unstructured grid, three-dimensional model based on the shallow water equations. *International Journal for Numerical Methods in Fluids*, 32(3), 331-348.

Chang, P.-H., Isobe, A., 2003. A numerical study on the Changjiang diluted water in the Yellow and East

China Seas. *Journal of Geophysical Research: Oceans*, 108(C9), 3299.

Chen, J., 1957. Notes on the development of the Yangtze Estuary. *Acta Geographica Sinica*, 23(3), 241-253. (in Chinese)

Chen, J., Zhu, H., Dong, Y., Sun, J., 1985. Development of the Changjiang estuary and its submerged delta. *Continental Shelf Research*, 4(1-2), 47-56.

Chen, J., Li, D., Chen, B., Hu, F., Zhu, H., Liu, C., 1999. The processes of dynamic sedimentation in the Changjiang Estuary. *Journal of Sea Research*, 41(1-2), 129-140.

Chen, J., Li, D., Chen, S., 2001a. Progress of estuarine research in China over last 50 years. *Science in China Series B: Chemistry*, 44(1), 1-9.

Chen, S.-L., Zhang, G.-A., Yang, S.-L., Shi, J.-Z., 2006. Temporal variations of fine suspended sediment concentration in the Changjiang River estuary and adjacent coastal waters, China. *Journal of Hydrology*, 331(1-2), 137-145.

Chen, S., Eisma, D., Kalf, J., 1994. In situ distribution of suspended matter during the tidal cycle in the elbe estuary. *Netherlands Journal of Sea Research*, 32(1), 37-48.

Chen, Z., Lin, B., 2000. Experimental study on the use of salt water to prevent siltation caused by density flow. *Journal of Sediment Research*, 5, 32-36. (in Chinese)

Chen, Z., Li, J., Shen, H., Zhanghua, W., 2001b. Yangtze River of China: historical analysis of discharge variability and sediment flux. *Geomorphology*, 41(2-3), 77-91.

Cheng, H., Jiang, Z., Shi, Z., 2003. Approximate estimations of threshold velocities for Non-uniform fine sediments at the South Passage of Changjiang Estuary, China. *Journal of Sediment Research*, 5, 37-40. (in Chinese)

Cheng, P., Wilson, R., 2008. Modeling Sediment Suspensions in an Idealized Tidal Embayment: Importance of Tidal Asymmetry and Settling Lag. *Estuaries and Coasts*, 31(5), 828-842.

Cheng, P., Wilson, R.E., Chant, R.J., Fugate, D.C., Flood, R.D., 2009. Modeling Influence of Stratification on Lateral Circulation in a Stratified Estuary. *Journal of Physical Oceanography*, 39(9), 2324-2337.

Cheng, P., Valle-Levinson, A., de Swart, H.E., 2011. A numerical study of residual circulation induced by asymmetric tidal mixing in tidally dominated estuaries. *Journal of Geophysical Research: Oceans*, 116(C1), C01017.

Cheng, R.T., Ling, C.-H., Gartner, J.W., Wang, P.F., 1999. Estimates of bottom roughness length and bottom shear stress in South San Francisco Bay, California. *Journal of Geophysical Research: Oceans*, 104(C4), 7715-7728.

Chien, N., Wan, Z., 1999. *Mechanics of sediment transport*. American Society of Civil Engineers, Reston, VA (US), 936 pp.

Chu, A., Wang, Z.B., Vriend, H.J.D., Stive, M., 2010. *A process-based approach to sediment transport in the Yangtze Estuary*. In: Smith, J.M., Lynett, P. (Eds.), Proceedings of the International Conference on Coastal Engineering. Coastal Engineering Research Council, Shanghai, 12 pp.

Corbett, D.R., McKee, B., Duncan, D., 2004. An evaluation of mobile mud dynamics in the Mississippi River deltaic region. *Marine Geology*, 209(1-4), 91-112.

Costa, R.C.F.G., 1989. *Flow-fine sediment hysteresis in sediment-stratified coastal water*. MSc thesis, University of Florida, Gainesville, 175 pp.

Cudaback, C.N., Jay, D.A., 2000. Tidal asymmetry in an estuarine pycnocline: Depth and thickness. *Journal of Geophysical Research: Oceans*, 105(C11), 26237-26251.

Cudaback, C.N., Jay, D.A., 2001. Tidal asymmetry in an estuarine pycnocline: 2. Transport. *Journal of Geophysical Research: Oceans*, 106(C2), 2639-2652.

CWRC, 2013. *Annual bulletin of Yangtze sediment*. Changjiang Water Resources Committee, Ministry of Water Resources. Changjiang Press, Wuhan, China, pp 47. (in Chinese)

Dai, Z., Liu, J.T., Fu, G., Xie, H., 2013. A thirteen-year record of bathymetric changes in the North Passage, Changjiang (Yangtze) estuary. *Geomorphology*, 187(0), 101-107.

Dalrymple, R.A., Trowbridge, J.H., Yue, D.K.P., Bentley, S.J., Kineke, G.C., Y., L., Mei, C.C., Shen, L., Traykovski, P.A., 2008. *Mechanisms of Fluid-Mud Interactions Under Waves*, Dept of Civil Engineering, The Johns Hopkins University, Baltimore, 11 pp.

Dastgheib, A., Roelvink, J.A., Wang, Z.B., 2008. Long-term process-based morphological modeling of the Marsdiep Tidal Basin. *Marine Geology*, 256(1-4), 90-100.

Dearnaley, M.P., 1996. Direct measurements of settling velocities in the owen tube: A comparison with gravimetric analysis. *Journal of Sea Research*, 36(1-2), 41-47.

de Nijs, M.A.J., 2012. *On sedimentation processes in a stratified estuarine system*. PhD thesis, Delft University of Technology, Delft, 302 pp.

de Nijs, M.A.J., Pietrzak, J.D., 2012. Saltwater intrusion and ETM dynamics in a tidally-energetic stratified estuary. *Ocean Modelling*, 49-50(0), 60-85.

de Stadler, M.B., Sarkar, S., Brucker, K.A., 2010. Effect of the Prandtl number on a stratified turbulent wake. *Physics of Fluids (1994-present)*, 22(9), 09510201-09510215.

de Vriend, H.J., 1991. Mathematical modelling and large-scale coastal behaviour. *Journal of Hydraulic Research*, 29(6), 727-740.

de Wit, P.J., 1995. *Liquefaction of Cohesive Sediments caused by Waves*. PhD thesis, Delft University of Technology, Delft, the Netherlands, 197 pp.

Dearnaley, M.P., 1996. Direct measurements of settling velocities in the owen tube: A comparison with gravimetric analysis. *Journal of Sea Research*, 36(1-2), 41-47.

Deleersnijder, E., Campin, J.-M., Delhez, E.J.M., 2001. The concept of age in marine modelling: I. Theory and preliminary model results. *Journal of Marine Systems*, 28(3-4), 229-267.

Deltares, 2014. *User Manual Delft3D-FLOW, version: 3.15*. Deltares, Delft, the Netherlands, 676 pp.

DHI, 2009. *Mike 3 coastal hydraulics and oceanography, user guide*. In: DHI (Ed.), Hørsholm, Denmark, 286 pp.

Dronkers, J., 1986. Tide-induced residual transport of fine sediment, *Physics of Shallow Estuaries and Bays*, Lecture Notes. Coastal Estuarine Study AGU, Washington, DC, 228-244.

Dyer, K.R., New, A.L., 1986. Intermittency in estuarine mixing. In: D.A. Wolfe (Ed.), *Estuarine Variability*. Academic Press, INC., 321-339.

Dyer, K.R., Cornelisse, J., Dearnaley, M.P., Fennessy, M.J., Jones, S.E., Kappenberg, J., McCave, I.N., Pejrup, M., Puls, W., Van Leussen, W., Wolfstein, K., 1996. A comparison of in situ techniques for estuarine floc settling velocity measurements. *Journal of Sea Research*, 36(1-2), 15-29.

Dyer, K.R., Manning, A.J., 1999. Observation of the size, settling velocity and effective density of flocs, and their fractal dimensions. *Journal of Sea Research*, 41(1-2), 87-95.

Edmonds, D.A., Slingerland, R.L., 2007. Mechanics of river mouth bar formation: Implications for the morphodynamics of delta distributary networks. *Journal of Geophysical Research: Earth Surface*, 112(F2), F02034.

Eisma, D., Bale, A.J., Dearnaley, M.P., Fennessy, M.J., van Leussen, W., Maldiney, M.A., Pfeiffer, A., Wells, J.T., 1996. Intercomparison of in situ suspended matter (floc) size measurements. *Journal of Sea Research*, 36(1-2), 3-14.

Fathi-Moghadam, M., Arman, A., Emamgholizadeh, S., Alikhani, A., 2011. Settling Properties of Cohesive Sediments in Lakes and Reservoirs. *Journal of Waterway, Port, Coastal, and Ocean Engineering*, 137(4), 204-209.

Feng, S., Cheng, R.T., Pangen, X., 1986. On tide-induced lagrangian residual current and residual transport: 1. Lagrangian residual current. *Water Resource Research*, 22(12), 1623-1634.

Fennessy, M.J., Dyer, K.R., Huntley, D.A., 1994. inssev: An instrument to measure the size and settling velocity of flocs in situ. *Marine Geology*, 117(1-4), 107-117.

Fringer, O.B., Gerritsen, M., Street, R.L., 2006. An unstructured-grid, finite-volume, nonhydrostatic, parallel coastal ocean simulator. *Ocean Modelling*, 14(3-4), 139-173.

Fuhrman, D.R., Fredsoe, J., Sumer, B.M., 2009. Bed slope effects on turbulent wave boundary layers: 1. Model validation and quantification of rough-turbulent results. *Journal of Geophysical Research*, 114(C3), C03024.

Gao, M., 2008. *A Pre-feasibility Study of Navigation Channel Regulation Works in North Channel, Yangtze Estuary*. MSc thesis, UNESCO-IHE, Delft, the Netherlands, 163 pp.

Gao, S., 2007. Modeling the growth limit of the Changjiang Delta. *Geomorphology*, 85(3-4), 225-236.

Gebhart, B., Jaluria, Y., Mahajan, R.L., Sammakia, B., 1988. *Buoyancy-Induced Flows and Transport* Hemisphere Pub. Corp., New York, 1001 pp.

Gerz, T., Schumann, U., 1996. A possible explanation of countergradient fluxes in homogeneous turbulence. *Theoretical and Computational Fluid Dynamics*, 8(3), 169-181.

Geyer, W.R., Smith, J.D., 1987. Shear Instability in a Highly Stratified Estuary. *Journal of Physical Oceanography*, 17(10), 1668-1679.

Geyer, W.R., Farmer, D.M., 1989. Tide-Induced Variation of the Dynamics of a Salt Wedge Estuary. *Journal of Physical Oceanography*, 19(8), 1060-1072.

Geyer, W.R., 1993. The importance of suppression of turbulence by stratification on the estuarine turbidity maximum. *Estuaries*, 16(1), 113-125.

Geyer, W.R., Beardsley, R.C., Lentz, S.J., Candela, J., Limeburner, R., Johns, W.E., Castro, B.M., Dias Soares, I., 1996. Physical oceanography of the Amazon shelf. *Continental Shelf Research*, 16(5–6), 575-616.

Geyer, W.R., Scully, M., Ralston, D., 2008. Quantifying vertical mixing in estuaries. *Environmental Fluid Mechanics*, 8(5-6), 495-509.

Gratiot, N., Michallet, H., Mory, M., 2005. On the determination of the settling flux of cohesive sediments in a turbulent fluid. *Journal of Geophysical Research: Oceans*, 110(C6), C06004.

157

Guan, W., 2003. *Transport and deposition of high-concentration suspensions of cohesive sediment in a macrotidal estuary*. PhD thesis, Hong Kong University of Science and Technology, Hong Kong, 185 pp.

Guan, W.B., Kot, S., Wolanski, E., 2005. 3-D fluid-mud dynamics in the Jiaojiang Estuary, China. *Estuarine, Coastal and Shelf Science*, 65(4), 747-762.

Guo, L., He, Q., 2011. Freshwater flocculation of suspended sediments in the Yangtze River, China. *Ocean Dynamics*, 61(2), 371-386.

Guo, L., van der Wegen, M., Roelvink, J.A., He, Q., 2014. The role of river flow and tidal asymmetry on 1-D estuarine morphodynamics. *Journal of Geophysical Research: Earth Surface*, 119(11), 2014JF003110.

Gust, G., Walger, E., 1976. The influence of suspended cohesive sediments on boundary-layer structure and erosive activity of turbulent seawater flow. *Marine Geology*, 22(3), 189-206.

Han, N., Lu, Z., 1983. *Settling properties of the sediments of the Changjiang Estuary in salt water*. In: A.O. Sinica (Ed.), Proceedings of international symposium on sedimentation on the continental shelf, with special reference to the East China Sea, Hangzhou, China, 483-493.

Hansen, D.V., Rattray, M., 1966. New dimension in estuary classification. *Limnology and Oceanography*, 11(3), 319-326.

Haralambidou, K., Sylaios, G., Tsihrintzis, V.A., 2010. Salt-wedge propagation in a Mediterranean micro-tidal river mouth. *Estuarine, Coastal and Shelf Science*, 90(4), 174-184.

Hattori, H., Morita, A., Nagano, Y., 2006. Nonlinear eddy diffusivity models reflecting buoyancy effect for wall-shear flows and heat transfer. *International Journal of Heat and Fluid Flow*, 27(4), 671-683.

Hsu, T.J., Traykovski, P.A., Kineke, G.C., 2007. On modeling boundary layer and gravity-driven fluid mud transport. *Journal of Geophysical Research*, 112, C04011.

Hu, D., Zhang, H., Zhong, D., 2009a. Properties of the Eulerian–Lagrangian method using linear interpolators in a three-dimensional shallow water model using z-level coordinates. *International Journal of Computational Fluid Dynamics*, 23(3), 271-284.

Hu, K., Ding, P., Wang, Z., Yang, S., 2009b. A 2D/3D hydrodynamic and sediment transport model for the Yangtze Estuary, China. *Journal of Marine Systems*, 77(1-2), 114-136.

Huang, J., 1981. Experimental study of settling properties of cohesive sediment in still water. *Journal of Sediment Research*, 6, 30-41. (in Chinese)

Huang, W., 2010. Enhancement of a Turbulence Sub-model for More Accurate Predictions of Vertical Stratifications in 3D Coastal and Estuarine Modeling. *The International Journal of Ocean and Climate Systems*, 1(1), 37-50.

Itsweire, E.C., Helland, K.N., Van Atta, C.W., 1986. The evolution of grid-generated turbulence in a stably stratified fluid. *Journal of Fluid Mechanics*, 162, 299-338.

Ivey, G.N., Winters, K.B., Koseff, J.R., 2008. Density Stratification, Turbulence, but How Much Mixing? *Annual Review of Fluid Mechanics*, 40(1), 169-184.

Jay, D.A., Smith, J.D., 1990. Residual circulation in shallow estuaries: 1. Highly stratified, narrow estuaries. *Journal of Geophysical Research: Oceans*, 95(C1), 711-731.

Jay, D.A., Musiak, J.D., 1994. Particle trapping in estuarine tidal flows. *Journal of Geophysical Research: Oceans*, 99(C10), 20445-20461.

Jay, D.A., Musiak, J.D., 1996. Internal tidal asymmetry in channel flows: Origins and consequences, *Mixing in Estuaries and Coastal Seas*. Coastal Estuarine Stud. AGU, Washington, DC, 211-249.

Jensen, J.H., Fredsoe, J., 2001. Sediment transport and backfilling of trenches in oscillatory flow. *Journal of Waterway Port Coastal and Ocean Engineering-Asce*, 127(5), 272-281.

Jiang, C., Swart, H., Li, J., Liu, G., 2013a. Mechanisms of along-channel sediment transport in the North Passage of the Yangtze Estuary and their response to large-scale interventions. *Ocean Dynamics*, 63(2-3), 283-305.

Jiang, G., Yao, Y., Tang, Z., 2002. The analysis for influencing factors of fine sediment flocculation in the Changjiang Estuary. *Acta Oceanologica Sinica*, 24(4), 51-57. (in Chinese)

Jiang, X., Lu, B., He, Y., 2013b. Response of the turbidity maximum zone to fluctuations in sediment discharge from river to estuary in the Changjiang Estuary (China). *Estuarine, Coastal and Shelf Science*, 131(0), 24-30.

Jing, L., Ridd, P.V., 1996. Wave-current bottom shear stresses and sediment resuspension in Cleveland Bay, Australia. *Coastal Engineering*, 29(1-2), 169-186.

Jones, J.H., 1973. Vertical Mixing in the Equatorial Undercurrent. *Journal of Physical Oceanography*, 3(3), 286-296.

Jung, K.T., Youll Jin, J., Kang, H.-W., Lee, H.J., 2004. An analytical solution for the local suspended sediment concentration profile in tidal sea region. *Estuarine, Coastal and Shelf Science*, 61(4),

657-667.

Karimpour, F., Venayagamoorthy, S.K., 2014. A simple turbulence model for stably stratified wall-bounded flows. *Journal of Geophysical Research: Oceans*, 119(2), 870-880.

Kays, W.M., 1994. Turbulent Prandtl Number-Where Are We? *Journal of Heat Transfer*, 116(2), 284-295.

Keller, K.H., Van Atta, C.W., 2000. Rapid vertical sampling in stably stratified turbulent shear flows. *Dynamics of Atmospheres and Oceans*, 31(1-4), 23-45.

Kineke, G.C., Sternberg, R.W., 1989. The effect of particle settling velocity on computed suspended sediment concentration profiles. *Marine Geology*, 90(3), 159-174.

Kineke, G.C., Sternberg, R.W., 1995. Distribution of fluid muds on the Amazon continental shelf. *Marine Geology*, 125(3-4), 193-233.

Kineke, G.C., Sternberg, R.W., Trowbridge, J.H., Geyer, W.R., 1995. Fluid-mud processes on the Amazon continental shelf. *Continental Shelf Research*, 16(5-6), 667-696.

King, J., Nickling, W.G., Gillies, J.A., 2008. Investigations of the law-of-the-wall over sparse roughness elements. *Journal of Geophysical Research: Earth Surface*, 113(F2), F02S07.

Kirby, R., Parker, W.R., 1983. Distribution and Behavior of Fine Sediment in the Severn Estuary and Inner Bristol Channel, U.K. *Canadian Journal of Fisheries and Aquatic Sciences*, 40(S1), s83-s95.

Kleinhans, M.G., Ferguson, R.I., Lane, S.N., Hardy, R.J., 2013. Splitting rivers at their seams: bifurcations and avulsion. *Earth Surface Processes and Landforms*, 38(1), 47-61.

Komori, S., Nagata, K., 1996. Effects of molecular diffusivities on counter-gradient scalar and momentum transfer in strongly stable stratification. *Journal of Fluid Mechanics*, 326, 205-237.

Krishnappan, B.G., Marsalek, J., Exall, K., Stephens, R. P., Rochfort, Q. and Seto, P., 2004. A Water Elutriation Apparatus for Measuring Settling Velocity Distribution of Suspended Solids in Combined Sewer Overflows. *Water Quality Research Journal of Canada*, 39(4), 432-438.

Krone, R.B., 1962. *Flume Studies of the Transport of Sediment in Estuarial Shoaling Processes*. Hydraulic Engineering Laboratory and Sanitary Engineering Research Laboratory, University of California, Berkeley, 110 pp.

L.Brennan, M., H.Schoellhamer, D., R.Burau, J., G.Monismith, S., 2002. Tidal asymmetry and variability of bed shear stress and sediment bed flux at a site in San Francisco Bay, USA. In: Winterwerp, J.C., Kranenburg, C. (Eds.), *Proceedings in Marine Science*. Elsevier, 93-107.

Lau, Y.L., 1994. Temperature effect on settling velocity and deposition of cohesive sediments. *Journal of Hydraulic Research*, 32(1), 41-51.

Le Hir, P., Bassoullet, P., Jestin, H., 2000. Application of the continuous modeling concept to simulate high-concentration suspended sediment in a macrotidal estuary. In: William, H.M., Mehta, A.J. (Eds.), *Proceedings in Marine Science*. Elsevier, 229-247.

Le Hir, P., Cayocca, F., 2002. 3D application of the continuous modelling concept to mud slides in open seas. In: Winterwerp, J.C., Kranenburg, C. (Eds.), *Proceedings in Marine Science*. Elsevier, 545-562.

Le Hir, P., Cayocca, F., Waeles, B., 2011. Dynamics of sand and mud mixtures: A multiprocess-based modelling strategy. *Continental Shelf Research*, 31(10, Supplement), S135-S149.

Lehfeldt, R., Bloss, S., 1988. Algebraic Turbulence Model for Stratified Tidal Flows. In: Dronkers, J., van Leussen, W. (Eds.), *Physical Processes in Estuaries*. Springer Berlin Heidelberg, 278-291.

Lesser, G.R., Roelvink, J.A., van Kester, J.A.T.M., Stelling, G.S., 2004. Development and validation of a three-dimensional morphological model. *Coastal Engineering*, 51(8-9), 883-915.

Levinton, J.S., Waldman, J.R., 2006. *The Hudson River Estuary*. Cambridge University Press, UK, 488 pp.

Li, J.F., Wan, X.N., He, Q., Ying, M., Shi, L.Q., Hutchinson, S.M., 2004. In-situ Observation of Fluid Mud in the North Passage of Yangtze Estuary, China. *China Ocean Engineering*, 18(1), 149-156.

Li, L., Zhu, J., Wu, H., 2012. Impacts of wind stress on saltwater intrusion in the Yangtze Estuary. *Science China Earth Sciences*, 55(7), 1178-1192.

Li, M., Zhong, L., Boicourt, W.C., 2005. Simulations of Chesapeake Bay estuary: Sensitivity to turbulence mixing parameterizations and comparison with observations. *Journal of Geophysical Research: Oceans*, 110(C12), C12004.

Li, M., Zhong, L., 2009. Flood–ebb and spring–neap variations of mixing, stratification and circulation in Chesapeake Bay. *Continental Shelf Research*, 29(1), 4-14.

Li, M.Z., Gust, 2000. Boundary layer dynamics and drag reduction in flows of high cohesive sediment suspensions. *Sedimentology*, 47(1), 71-86.

Li, W., Cheng, H., Li, J., Dong, P., 2008. Temporal and spatial changes of dunes in the Changjiang (Yangtze) estuary, China. *Estuarine, Coastal and Shelf Science*, 77(1), 169-174.

Li, Y., Mehta, A.J., 1998. Assessment of Hindered Settling of Fluid Mudlike Suspensions. *Journal of*

Hydraulic Engineering, 124(2), 176-178.

Li, Y., Mehta, A.J., 2000. Fluid mud in the wave-dominated environment revisited. In: William, H.M., Mehta, A.J. (Eds.), *Proceedings in Marine Science*. Elsevier, 79-93.

Liu, G., Zhu, J., Wang, Y., Wu, H., Wu, J., 2011. Tripod measured residual currents and sediment flux: Impacts on the silting of the Deepwater Navigation Channel in the Changjiang Estuary. *Estuarine, Coastal and Shelf Science*, 93, 192-201.

Liu, G., Wu, H., Guo, W., Zhu, J., Sun, L., 2012. Dispersal and fate of dredged materials disposed of in the Changjiang Estuary determined by use an in situ rare earth element tracer. *China Ocean Engineering*, 25, 495-506.

Ma, C., Wu, D., Lin, X., Yang, J., Ju, X., 2010. An open-ocean forcing in the East China and Yellow seas. *Journal of Geophysical Research: Oceans*, 115(C12), C12056.

Ma, G., Shi, F., Liu, S., Qi, D., 2011. Hydrodynamic modeling of Changjiang Estuary: Model skill assessment and large-scale structure impacts. *Applied Ocean Research*, 33(1), 69-78.

Madsen, O.S., Wood, W., 2002. *Sediment Transport Outside the Surf Zone*. In: W. T. (Ed.), Coastal Engineering Manual. Part III Coastal Processes. U.S. Army Corps of Engineers, Washington DC, 72 pp.

Maggi, F., 2007. Variable fractal dimension: A major control for floc structure and flocculation kinematics of suspended cohesive sediment. *Journal of Geophysical Research: Oceans*, 112(C7), C07012.

Manning, A.J., Langston, W.J., Jonas, P.J., 2010. A review of sediment dynamics in the Severn Estuary: influence of flocculation. *Marine Pollution Bulletin*, 61(1-3), 37-51.

Manning, A.J., Schoellhamer, D.H., 2013. Factors controlling floc settling velocity along a longitudinal estuarine transect. *Marine Geology*, 345(0), 266-280.

Mantovanelli, A., 2005. *A new approach for measuring in situ the concentration and settling velocity of suspended cohesive sediment*. PhD thesis, James Cook University, Queensland, Australia, 190 pp.

Mao, Z., Pan, D., Shen, H., 2001. The Patterns of transportation and morpholpgical features of deposition of suspended sediment in the Changjiang Estuary. *Geographical Research*, 20(2), 170-177. (in Chinese)

Maren, D., Yang, S.-L., He, Q., 2013. The impact of silt trapping in large reservoirs on downstream morphology: the Yangtze River. *Ocean Dynamics*, 63(6), 691-707.

Markussen, T.N., Andersen, T.J., 2013. A simple method for calculating in situ floc settling velocities based on effective density functions. *Marine Geology*, 344(0), 10-18.

Matsumoto, K., Takanezawa, T., Ooe, M., 2000. Ocean Tide Models Developed by Assimilating TOPEX/POSEIDON Altimeter Data into Hydrodynamical Model: A Global Model and a Regional Model around Japan. *Journal of Oceanography*, 56(5), 567-581.

McAnally, W.H., 2000. *Aggregation and deposition of estuarial fine sediment*, US Army Engineer Research and Development Center, Vicksburg, MS, 366 pp.

McAnally, W.H., Friedrichs, C., Hamilton, D., Hayter, E., Shrestha, P., Rodriguez, H., Scheremet, A., Teeter, A., 2007a. Management of Fluid Mud in Estuaries, Bays, and Lakes. I Present State of Understanding on Character and Behavior. *Journal of Hydraulic Engineering*(January), 9-22.

McAnally, W.H., Teeter, A., Schoellhamer, D., Friedrichs, C., Hamilton, D., Hayter, E., Shrestha, P., Rodriguez, H., Scheremet, A., Kirby, R., 2007b. Management of Fluid Mud in Estuaries, Bays, and Lakes. II Measurement, Modeling, and Management. *Journal of Hydraulic Engineering* (January), 23-38.

McLaughlin, R.T., 1961. Settling Properties of Suspensions. *Transactions of the American Society of Civil Engineers*, 126(1), 1780-1786.

Mehta, A.J., 1991. Understanding fluid mud in a dynamic environment. *Geo-Marine Letters*, 11(3-4), 113-118.

Mehta, A.J., 1989. On estuarine cohesive sediment suspension behavior. *Journal of Geophysical Research: Oceans*, 94(C10), 14303-14314.

Mehta, A.J., McAnally, W.H., 2008. Sedimentation Engineering: Processes, Measurements, Modeling, and Practice, In: Garcia, M.H. (Eds.) *Fine-grained sediment transport*. American Society of Civil Engineers, Virginia, 253-306.

Mehta, A.J., 2014. *An Introduction to Hydraulics of Fine Sediment Transport* Advanced Series on Ocean Engineering 38. World Scientific Publishing Company, 1039 pp.

Meire, P., Ysebaert, T., Damme, S., Bergh, E., Maris, T., Struyf, E., 2005. The Scheldt estuary: a description of a changing ecosystem. *Hydrobiologia*, 540(1-3), 1-11.

Mellor, G.L., Yamada, T., 1982. Development of a turbulence closure model for geophysical fluid problems. *Rev. Geophys.*, 20(4), 851-875.

Mellor, G.L., 2001. One-Dimensional, Ocean Surface Layer Modeling: A Problem and a Solution. *Journal of Physical Oceanography*, 31(3), 790-809.

Middelburg, J., Nieuwenhuize, J., Iversen, N., Høgh, N., de Wilde, H., Helder, W., Seifert, R., Christof, O., 2002. Methane distribution in European tidal estuaries. *Biogeochemistry*, 59(1-2), 95-119.

Mikkelsen, O.A., Hill, P.S., Milligan, T.G., 2007. Seasonal and spatial variation of floc size, settling velocity, and density on the inner Adriatic Shelf (Italy). *Continental Shelf Research*, 27(3–4), 417-430.

Miles, J., 1986. Richardson's criterion for the stability of stratified shear flow. *Physics of Fluids (1958-1988)*, 29(10), 3470-3471.

Mitchell, S.B., Burgess, H.M., Pope, D.J., 2006. Stratification and fine sediment transport mechanisms in a semi-enclosed tidal lagoon (Pagham Harbour, West Sussex). *Water and Environment Journal*, 20(4), 248-255.

Monti, P., Fernando, H.J.S., Princevac, M., Chan, W.C., Kowalewski, T.A., Pardyjak, E.R., 2002. Observations of Flow and Turbulence in the Nocturnal Boundary Layer over a Slope. *Journal of the Atmospheric Sciences*, 59(17), 2513-2534.

Munk, W.H., Anderson, E.R., 1948. Notes on the theory of the thermocline. *Journal of Marine Research*, 7(3), 276-295.

Murakami, S., Kato, S., Chikamoto, T., Laurence, D., Blay, D., 1996. New low-Reynolds-number k-ε model including damping effect due to buoyancy in a stratified flow field. *International Journal of Heat and Mass Transfer*, 39(16), 3483-3496.

Nepf, H.M., Geyer, W.R., 1996. Intratidal variations in stratification and mixing in the Hudson estuary. *Journal of Geophysical Research: Oceans*, 101(C5), 12079-12086.

Nguyen, A.D., 2008. *Salt Intrusion, Tides and Mixing in Multi-channel Estuaries*. PhD thesis, Delft, 174 pp.

Nunes Vaz, R.A., Simpson, J.H., 1994. Turbulence closure modeling of estuarine stratification. *Journal of Geophysical Research: Oceans*, 99(C8), 16143-16160.

Ohya, Y., Neff, D.E., Meroney, R.N., 1997. Turbulence structure in a stratified boundary layer under stable conditions. *Boundary-layer Meteorology*, 83(1), 139-162.

Oliveira, A., Baptista, A.M., 1995. A comparison of integration and interpolation Eulerian-Lagrangian methods. *International Journal for Numerical Methods in Fluids*, 21(3), 183-204.

Owen, M.W., 1971. *The effects of turbulence on the settling velocity of silt flocs*, Proceedings 14th Congress International Association Hydraulic Research, Paris, D4-1-D4-6.

Owen, M.W., 1972. *Effect of Temperature on the Settling Velocities of an Estuary Mud*, Hydraulics Research Station, Wallingford, HR Report No. INT 106.

Owen, R.B., Zozulya, A.A., 2000. In-line digital holographic sensor for monitoring and characterizing marine particulates. *Optical Engineering*, 39(8), 2187-2197.

Pacanowski, R.C., Philander, S.G.H., 1981. Parameterization of Vertical Mixing in Numerical Models of Tropical Oceans. *Journal of Physical Oceanography*, 11(11), 1443-1451.

Partheniades, E., 1965. Erosion and Deposition of Cohesive Soils. *Journal of the Hydraulics Division, ASCE*, 91(1), 105-139.

PDC, 1986. *Regulation of the Yangtze Estuary (General report): Review on investigations and studies carried out by ECIDI-SB*, Port and Delta Consortium B.V., Papendrecht, the Netherlands, 74 pp.

Pejrup, M., Mikkelsen, O.A., 2010. Factors controlling the field settling velocity of cohesive sediment in estuaries. *Estuarine, Coastal and Shelf Science*, 87(2), 177-185.

Penland, S., Suter, J.R., 1989. The geomorphology of the Mississippi River chenier plain. *Marine Geology*, 90(4), 231-258.

Peters, H., Gregg, M.C., Toole, J.M., 1988. On the parameterization of equatorial turbulence. *Journal of Geophysical Research: Oceans*, 93(C2), 1199-1218.

Peters, H., 1997. Observations of Stratified Turbulent Mixing in an Estuary: Neap-to-spring Variations During High River Flow. *Estuarine, Coastal and Shelf Science*, 45(1), 69-88.

Postma, H., 1961. Transport and accumulation of suspended matter in the Dutch Wadden Sea. *Netherlands Journal of Sea Research*, 1(1-2), 148-190.

Pritchard, D., 2005. Suspended sediment transport along an idealised tidal embayment: settling lag, residual transport and the interpretation of tidal signals. *Ocean Dynamics*, 55(2), 124-136.

Puig, P., Ogston, A.S., Mullenbach, B.L., Nittrouer, C.A., Parsons, J.D., Sternberg, R.W., 2004. Storm-induced sediment gravity flows at the head of the Eel submarine canyon, northern California margin. *Journal of Geophysical Research*, 109(C03019), 10.

Puls, W., Kuehl, H., Heymann, K., 1988. Settling Velocity of Mud Flocs: Results of Field Measurements in the Elbe and the Weser Estuary. In: Dronkers, J., van Leussen, W. (Eds.), *Physical Processes in Estuaries*. Springer Berlin Heidelberg, 404-424.

Qi, D., 2007. *Development a numerical modeling platform for the Yangtze Estuary waterway and Its application*, Estuarine and Coastal Science Research Center (ECSRC), Shanghai, 144 pp. (in Chinese)

Qi, D., Ma, G., Gu, F., Mou, L., 2010. An unstructured grid hydrodynamic and sediment transport model for Changjiang Estuary. *Journal of Hydrodynamics, Series B*, 22(5, Supplement 1), 1015-1021.

Qiu, B., Imasato, N., Awaji, T., 1988. Baroclinic instability of buoyancy-driven coastal density currents. *Journal of Geophysical Research: Oceans*, 93(C5), 5037-5050.

Rodi, W., 1993. *Turbulence models and their application in hydraulics: a state-of-the-art review*. Balkema, Rotterdam, 115 pp.

Roelvink, J.A., 2006. Coastal morphodynamic evolution techniques. *Coastal Engineering*, 53(2-3), 277-287.

Roelvink, J.A., Reniers, A.J.H.M., 2012. *A guide to modeling coastal morphology*. Advances in coastal and ocean engineering. World Scientific, Singapore, 274 pp.

Ross, M.A., 1988. *Vertical structure of estuarine fine sediment suspensions*. PhD thesis, University of Florida, Gainesville, 187 pp.

Ross, M.A., Mehta, A.J., 1989. On the mechanics of lutoclines and fluid mud. *Journal of Coastal Research*(Special Issue No. 5), 51-62.

Rouse, H., 1938. *Experiments on the mechanics of sediment suspension*. In: Den Hartog, J.P., Peters, H. (Eds.), Proceedings 5th International Congress on Applied Mechanics. Wiley, Cambridge, Massachusetts, 550-554.

Ruessink, B.G., Walstra, D.J.R., Southgate, H.N., 2003. Calibration and verification of a parametric wave model on barred beaches. *Coastal Engineering*, 48(3), 139-149.

Saeijs, H.L.F., 2008. *Turning the tide: Essays on Dutch ways with water*. VSSD, 142 pp.

Schumann, U., Gerz, T., 1995. Turbulent Mixing in Stably Stratified Shear Flows. *Journal of Applied Meteorology*, 34(1), 33-48.

Scully, M.E., Friedrichs, C.T., 2007. The Importance of Tidal and Lateral Asymmetries in Stratification to Residual Circulation in Partially Mixed Estuaries. *Journal of Physical Oceanography*, 37(6), 1496-1511.

Seybold, H., Andrade, J.S., Herrmann, H.J., 2007. Modeling river delta formation. *Proceedings of the National Academy of Sciences*, 104(43), 16804-16809.

Shao, Y., Yan, Y., Maa, J., 2011. In Situ Measurements of Settling Velocity near Baimao Shoal in Changjiang Estuary. *Journal of Hydraulic Engineering*, 137(3), 372-380.

Shen, H., Zhang, C., 1992. Mixing of salt water and fresh water in the Changjiang River estuary and its effects on suspended sediment. *Chinese Geographical Science*, 2(4), 373-381. (in Chinese)

Shen, G., Xie, Z., 2004. Three Gorges Project: Chance and Challenge. *Science*, 304(5671), 681.

Shi, B.W., Yang, S.L., Wang, Y.P., Bouma, T.J., Zhu, Q., 2012. Relating accretion and erosion at an exposed tidal wetland to the bottom shear stress of combined current–wave action. *Geomorphology*, 138(1), 380-389.

Shi, J.Z., 1998. Acoustic observations of fluid mud and interfacial waves, Hangzhou Bay, China. *Journal Coastal Research* 14, 1348-1353.

Shi, J.Z., 2010. Tidal resuspension and transport processes of fine sediment within the river plume in the partially-mixed Changjiang River estuary, China: A personal perspective. *Geomorphology*, 121(3-4), 133-151.

Shi, Y.-L., Yang, W., Ren, M.-E., 1985. Hydrological characteristics of the Changjiang and its relation to sediment transport to the sea. *Continental Shelf Research*, 4(1-2), 5-15.

Shivaprasad, A., Vinita, J., Revichandran, C., Reny, P.D., Deepak, M.P., Muraleedharan, K.R., Naveen Kumar, K.R., 2013. Seasonal stratification and property distributions in a tropical estuary (Cochin estuary, west coast, India). *Hydrology and Earth System Sciences*, 17(1), 187-199.

Simpson, J.H., Brown, J., Matthews, J., Allen, G., 1990. Tidal straining, density currents, and stirring in the control of estuarine stratification. *Estuaries*, 13(2), 125-132.

Slaa, S., He, Q., Maren, D.S., Winterwerp, J.C., 2013. Sedimentation processes in silt-rich sediment systems. *Ocean Dynamics*, 63(4), 399-421.

Smagorinsky, J., 1963. *General circulation experiments with the primitive equations*. Monthly Weather Review, 91, 99-164.

SOA, 2014. *Annuar Bulletin of the Chinese Sea Level*, State Oceanic Administration, Beijing, China, www.soa.gov.cn/zwgk/hygb/zghpmgb/201503/t20150318_36408.html. (in Chinese)

Song, D., Wang, X.H., 2013. Suspended sediment transport in the Deepwater Navigation Channel, Yangtze River Estuary, China, in the dry season 2009: 2. Numerical simulations. *Journal of Geophysical Research: Oceans*, 118(10), 5568-5590.

Song, D., Wang, X.H., Cao, Z., Guan, W., 2013. Suspended sediment transport in the Deepwater Navigation Channel, Yangtze River Estuary, China, in the dry season 2009: 1. Observations over spring and neap tidal cycles. *Journal of Geophysical Research: Oceans*, 118(10), 5555-5567.

Sorbjan, Z., 2006. Local Structure of Turbulence in Stably Stratified Boundary Layers. *Journal of the Atmospheric Sciences*, 63(5), 1526-1537.

Stacey, M.T., Monismith, S.G., Burau, J.R., 1999. Observations of Turbulence in a Partially Stratified Estuary. *Journal of Physical Oceanography*, 29(8), 1950-1970.

Stelling, G.S., Van Kester, J.A.T.M., 1994. On the approximation of horizontal gradients in sigma co-ordinates for bathymetry with steep bottom slopes. *International Journal for Numerical Methods in Fluids*, 18(10), 915-935.

Stelling, G.S., Duinmeijer, S.P.A., 2003. A staggered conservative scheme for every Froude number in rapidly varied shallow water flows. *International Journal for Numerical Methods in Fluids*, 43(12), 1329-1354.

Stillinger, D.C., Helland, K.N., Van Atta, C.W., 1983. Experiments on the transition of homogeneous turbulence to internal waves in a stratified fluid. *Journal of Fluid Mechanics*, 131, 91-122.

Strang, E.J., Fernando, H.J.S., 2001. Vertical Mixing and Transports through a Stratified Shear Layer. *Journal of Physical Oceanography*, 31(8), 2026-2048.

Su, J., Wang, K., 1989. Changjiang river plume and suspended sediment transport in Hangzhou Bay. *Continental Shelf Research*, 9(1), 93-111.

Sutherland, T.F.; Lane, P.M.; Amos, C.L. and Downing, J., 2000. Calibration of Optical Backscatter Sensors for Suspended Sediment of Varying Darkness Levels. *Marine Geology*, 162, 587-597.

Talke, S.A., de Swart, H.E., Schuttelaars, H.M., 2009. Feedback between residual circulations and sediment distribution in highly turbid estuaries: An analytical model. *Continental Shelf Research*, 29(1), 119-135.

Tambo, N., 1964. Settling property determination. *Memoirs of the Faculty of Engineering, Hokkaido University*, 11(5), 559-584.

Taylor, E., Leonard, J., 1990. *Sediment consolidation and permeability at the Barbados forearc*. In: J.C. Moore, Mascle, A., et al (Ed.), Proceedings of the Ocean Drilling Program, Scientific Results, Vol. 110. College Station, TX (Ocean Drilling Program), 289-308.

Teisson, C., Simonin, O., Galland, J.C., Laurence, D., 1992. Turbulence and mud sedimentation: a Reynolds stress model and a two-phase flow model. *Coastal Engineering Proceedings*(23), 2853-2866.

Traykovski, P., Geyer, W.R., Irish, J.D., Lynch, J.F., 2000. The role of wave-induced density-driven fluid mud flows for cross-shelf transport on the Eel River continental shelf. *Continental Shelf Research*, 20(16), 2113-2140.

Umlauf, L., Burchard, H., 2005. Second-order turbulence closure models for geophysical boundary layers. A review of recent work. *Continental Shelf Research*, 25(7–8), 795-827.

Umlauf, L., 2009. The Description of Mixing in Stratified Layers without Shear in Large-Scale Ocean Models. *Journal of Physical Oceanography*, 39(11), 3032-3039.

Uncles, R.J., Jordan, M.B., 1979. Residual fluxes of water and salt at two stations in the Severn Estuary. *Estuarine and Coastal Marine Science*, 9(3), 287-302.

Uncles, R.J., Ong, J.E., Gong, W.K., 1990. Observations and analysis of a stratification-destratification event in a tropical estuary. *Estuarine, Coastal and Shelf Science*, 31(5), 651-665.

Uncles, R.J., Stephens, J.A., Smith, R.E., 2002. The dependence of estuarine turbidity on tidal intrusion length, tidal range and residence time. *Continental Shelf Research*, 22(11-13), 1835-1856.

Uncles, R.J., Stephens, J.A., Law, D.J., 2006. Turbidity maximum in the macrotidal, highly turbid Humber Estuary, UK: Flocs, fluid mud, stationary suspensions and tidal bores. *Estuarine, Coastal and Shelf Science*, 67(1-2), 30-52.

Valle-Levinson, A. (Ed.), 2010. *Contemporary Issues in Estuarine Physics*. Cambridge University Press, UK, 326 pp.

van der Wegen, M., Jaffe, B.E., 2014. Processes governing decadal-scale depositional narrowing of the major tidal channel in San Pablo Bay, California, USA. *Journal of Geophysical Research: Earth Surface*, 119(5), 2013JF002824.

van Kessel, T., Kranenburg, C., 1998. Wave-induced liquefaction and flow of subaqueous mud layers. *Coastal Engineering*, 34, 109-127.

van Kessel, T., Vanlede, J., de Kok, J., 2011a. Development of a mud transport model for the Scheldt estuary. *Continental Shelf Research*, 31(10, Supplement), S165-S181.

van Kessel, T., Winterwerp, H., Van Prooijen, B., Van Ledden, M., Borst, W., 2011b. Modelling the seasonal dynamics of SPM with a simple algorithm for the buffering of fines in a sandy seabed. *Continental Shelf Research*, 31(10, Supplement), S124-S134.

van Leussen, W., 1988. Aggregation of Particles, Settling Velocity of Mud Flocs A Review. In: Dronkers, J., van Leussen, W. (Eds.), *Physical Processes in Estuaries*. Springer Berlin Heidelberg,

347-403.

van Leussen, W., 1994. *Estuarine macroflocs and their role in fine-grained sediment transport*. PhD thesis, Utrecht University, 488 pp.

van Leussen, W., 2011. Macroflocs, fine-grained sediment transports, and their longitudinal variations in the Ems Estuary. *Ocean Dynamics*, 61(2-3), 387-401.

van Maanen, B., Sottolichio, A., 2013. *Hydro- and sediment dynamics in the Gironde Estuary (France): model validation and sea level rise effects*, 7th International Conference on Coastal Dynamics, France, 1787-1797.

van Maren, D.S., Winterwerp, J.C., Wu, B.S., Zhou, J.J., 2009. Modelling hyperconcentrated flow in the Yellow River. *Earth Surface Processes and Landforms*, 34(4), 596-612.

van Maren, D.S., Winterwerp, J.C., 2013. The role of flow asymmetry and mud properties on tidal flat sedimentation. *Continental Shelf Research*, 60, Supplement(0), S71-S84.

van Rijn, L.C., 1993. *Principles of sediment transport in rivers, estuaries and coastal seas*. Aqua Publications, Amsterdam, the Netherlands, 700 pp.

van Rijn, L.C., 2007. Unified view of sediment transport by currents and waves. II: Suspended transport. *Journal of Hydraulic Engineering-Asce*, 133(6), 668-689.

Vasil'ev, O.F., Voropaeva, O.F., Kurbatskii, A.F., 2011. Turbulent mixing in stably stratified flows of the environment: The current state of the problem (Review). *Izvestiya, Atmospheric and Oceanic Physics*, 47(3), 265-280.

Venayagamoorthy, S.K., Stretch, D.D., 2010. On the turbulent Prandtl number in homogeneous stably stratified turbulence. *Journal of Fluid Mechanics*, 644, 359-369.

Vinzon, S.B., Mehta, A.J., 2003. Lutoclines in High Concentration Estuaries Some Observations at the Mouth of the Amazon. *Journal of Coastal Research*, 19(2), 243-253.

Wan, Y., Qi, D., 2009. *Preliminary Analysis on the Impact of Different Stages of Yangtze Estuary Deepwater Channel Regulation Project*, 3rd International Conference on Estuaries & Coasts, Sendai, Japan, 181-187.

Wan, Y., 2013. *Experimental study on settling velocity of fine sediments in the Yangtze Estuary*, Estuarine and Coastal Science Research Center, Shanghai, 158 pp. (in Chinese)

Wan, Y., Gu, F., Wu, H., Roelvink, D., 2014a. Hydrodynamic evolutions at the Yangtze Estuary from 1998 to 2009. *Applied Ocean Research*, 47(0), 291-302.

Wan, Y., Roelvink, D., Li, W., Qi, D., Gu, F., 2014b. Observation and modeling of the storm-induced fluid mud dynamics in a muddy-estuarine navigational channel. *Geomorphology*, 217(0), 23-36.

Wang, L., 2010. *Tide Driven Dynamics of Subaqueous Fluid Mud Layers in Turbidity Maximum Zones of German Estuaries*. PhD thesis, Universitaet Bremen, Bremen, German, 96 pp.

Wang, X.H., 2002. Tide-Induced Sediment Resuspension and the Bottom Boundary Layer in an Idealized Estuary with a Muddy Bed. *Journal of Physical Oceanography*, 32(11), 3113-3131.

Wang, Y., He, Q., 2007. The critical diameter for distinguishing bed material load and wash load of sediment in the Yangtze Estuary, China. In: Maa, J., Sanford, L., Schoellhamer, D. (Eds.), *Proceedings in Marine Science*. Elsevier, 277-289.

Wang, Y., Yu, Q., Gao, S., 2011. Relationship between bed shear stress and suspended sediment concentration: annular flume experiments. *International journal of sediment research*, 26(4), 513-523.

Wang, Y., Dong, P., Oguchi, T., Chen, S., Shen, H., 2013. Long-term (1842-2006) morphological change and equilibrium state of the Changjiang (Yangtze) Estuary, China. *Continental Shelf Research*, 56(0), 71-81.

Wang, Z., Larsen, P., Nestmann, F., Dittrich, A., 1998. Resistance and Drag Reduction of Flows of Clay Suspensions. *Journal of Hydraulic Engineering*, 124(1), 41-49.

Wang, Z.B., 1989. *Mathematical modelling of morphological processes in estuaries*. PhD thesis, Delft University of Technology, Delft, the Netherlands, 220 pp.

Wang, Z.B., Winterwerp, J.C., He, Q., 2014. Interaction between suspended sediment and tidal amplification in the Guadalquivir Estuary. *Ocean Dynamics*, 64(10), 1487-1498.

Warner, J., Butman, B., Dalyander, P., 2008. Storm-driven sediment transport in Massachusetts Bay. *Continental Shelf Research*, 28(2), 257-282.

Warner, J.C., Sherwood, C.R., Arango, H.G., Signell, R.P., 2005. Performance of four turbulence closure models implemented using a generic length scale method. *Ocean Modelling*, 8(1-2), 81-113.

Whitehead, J.A., 1987. On the ratio of the mixing coefficients of heat and salt of Antarctic Bottom Water in the North Atlantic. *Journal of Geophysical Research: Oceans*, 92(C3), 2981-2984.

Willmott, C.J., Ackleson, S.G., Davis, R.E., Feddema, J.J., Klink, K.M., Legates, D.R., O'Donnell, J., Rowe, C.M., 1985. Statistics for the evaluation and comparison of models. *Journal of Geophysical Research: Oceans*, 90(C5), 8995-9005.

Winterwerp, J.C., 1998. A simple model for turbulence induced flocculation of cohesive sediment. *Journal of Hydraulic Research*, 36(3), 309-326.

Winterwerp, J.C., 1999. *On the dynamics of high-concentrated mud suspensions*. PhD thesis, Delft University of Technology, Delft, the Netherlands, 204 pp.

Winterwerp, J.C., 2002. On the flocculation and settling velocity of estuarine mud. *Continental Shelf Research*, 22(9), 1339-1360.

Winterwerp, J.C., Bruens, A.W., Gratiot, N., Kranenburg, C., Mory, M., Toorman, E.A., 2002. Dynamics of Concentrated Benthic suspension layers. In: Winterwerp, J.C., Kranenburg, C. (Eds.), *Proceedings in Marine Science*. Elsevier, 41-55.

Winterwerp, J.C., van Kesteren, W.G.M., 2004. Introduction to the Physics of Cohesive Sediment in the Marine Environment. In: van Loon, A.J. (Ed.), *Developments in sedimentology 56*. Elsevier, Doorweth, the Netherlands, 576 pp.

Winterwerp, J.C., 2006. Stratification effects by fine suspended sediment at low, medium, and very high concentrations. *Journal of Geophysical Research*, 111(C5).

Winterwerp, J.C., Lely, M., He, Q., 2009. Sediment-induced buoyancy destruction and drag reduction in estuaries. *Ocean Dynamics*, 59(5), 781-791.

Winterwerp, J.C., 2011a. Fine sediment transport by tidal asymmetry in the high-concentrated Ems River: indications for a regime shift in response to channel deepening. *Ocean Dynamics*, 61(2-3), 203-215.

Winterwerp, J.C., 2011b. 2.15 - The Physical Analyses of Muddy Sedimentation Processes. In: Wolanski, E., McLusky, D. (Eds.), *Treatise on Estuarine and Coastal Science*. Academic Press, Waltham, 311-360.

Winterwerp, J.C., 2013. *On the response of tidal rivers to deepening and narrowing, Risks for a regime shift towards hyper-turbid conditions*, Deltares, Delft, the Netherlands, 83 pp.

Wolanski, E., Chappell, J., Ridd, P., Vertessy, R., 1988. Fluidization of mud in estuaries. *Journal of Geophysical Research*, 93(C3), 2351-2361.

Wolanski, E., 1995. Transport of sediment in mangrove swamps. *Hydrobiologia*, 295(1-3), 31-42.

Wolanski, E., 2007. *Estuarine Ecohydrology*. Elsevier, Amsterdam, the Netherlands, 157 pp.

Wu, H., Zhu, J., Ho Choi, B., 2010. Links between saltwater intrusion and subtidal circulation in the Changjiang Estuary: A model-guided study. *Continental Shelf Research*, 30(17), 1891-1905.

Wu, H., Zhu, J., Shen, J., Wang, H., 2011. Tidal modulation on the Changjiang River plume in summer. *Journal of Geophysical Research: Oceans*, 116(C8), C08017.

Wu, J., Huang, J., Han, X., Xie, Z., Gao, X., 2003. Three-Gorges Dam--Experiment in Habitat Fragmentation? *Science*, 300(5623), 1239-1240.

Wu, J., Liu, J.T., Wang, X., 2012. Sediment trapping of turbidity maxima in the Changjiang Estuary. *Marine Geology*, 303-306(0), 14-25.

Wu, W., Wang, S.S.Y., 2004. Depth-averaged 2-D calculation of tidal flow, salinity and cohesive sediment transport in estuaries. *International journal of sediment research*, 19(3), 172-190.

Wu, W., 2007. *Computational River Dynamics*. Taylor & Francis, UK, 494 pp.

Xu, F., Wang, D.-P., Riemer, N., 2010. An idealized model study of flocculation on sediment trapping in an estuarine turbidity maximum. *Continental Shelf Research*, 30(12), 1314-1323.

Xu, K., Milliman, J.D., 2009. Seasonal variations of sediment discharge from the Yangtze River before and after impoundment of the Three Gorges Dam. *Geomorphology*, 104(3-4), 276-283.

Yamazaki, H., Osborn, T., 1990. Dissipation estimates for stratified turbulence. *Journal of Geophysical Research: Oceans*, 95(C6), 9739-9744.

Yang, S.L., Milliman, J.D., Xu, K.H., Deng, B., Zhang, X.Y., Luo, X.X., 2014. Downstream sedimentary and geomorphic impacts of the Three Gorges Dam on the Yangtze River. *Earth-Science Reviews*, 138(0), 469-486.

You, Z.-J., 2004. The effect of suspended sediment concentration on the settling velocity of cohesive sediment in quiescent water. *Ocean Engineering*, 31(16), 1955-1965.

Yu, Q., Wang, Y., Gao, J., Gao, S., Flemming, B., 2014. Turbidity maximum formation in a well-mixed macrotidal estuary: The role of tidal pumping. *Journal of Geophysical Research: Oceans*, 119(11), 7705-7724.

Zeidan, M., Xu, B.H., Jia, X., Williams, R.A., 2007. Simulation of Aggregate Deformation and Breakup in Simple Shear Flows Using a Combined Continuum and Discrete Model. *Chemical Engineering Research and Design*, 85(12), 1645-1654.

Zhang, E., Savenije, H.H.G., Chen, S., Chen, J., 2012a. Water abstraction along the lower Yangtze River, China, and its impact on water discharge into the estuary. *Physics and Chemistry of the Earth, Parts A/B/C*, 47-48(0), 76-85.

Zhang, E.F., Savenije, H.H.G., Chen, S.L., Mao, X.H., 2012b. An analytical solution for tidal propagation in the Yangtze Estuary, China. *Hydrology and Earth System Sciences*, 16(9), 3327-3339.

Zhang, J.-F., Zhang, Q.-H., 2011. Lattice Boltzmann simulation of the flocculation process of cohesive sediment due to differential settling. *Continental Shelf Research*, 31(10, Supplement), S94-S105.

Zhang, Y., Baptista, A.M., 2008. SELFE: A semi-implicit Eulerian-Lagrangian finite-element model for cross-scale ocean circulation. *Ocean Modelling*, 21, 71-96.

Zhu, Q., Yang, S., Ma, Y., 2014. Intra-tidal sedimentary processes associated with combined wave-current action on an exposed, erosional mudflat, southeastern Yangtze River Delta, China. *Marine Geology*, 347(0), 95-106.

Zijlema, M., 2010. Computation of wind-wave spectra in coastal waters with SWAN on unstructured grids. *Coastal Engineering*, 57(3), 267-277.

Zilitinkevich, S., Esau, I., 2007. Similarity theory and calculation of turbulent fluxes at the surface for the stably stratified atmospheric boundary layer. *Boundary-layer Meteorology*, 125(2), 193-205.

Zilitinkevich, S.S., Elperin, T., Kleeorin, N., Rogachevskii, I., 2007. Energy- and flux-budget (EFB) turbulence closure model for stably stratified flows. Part I: steady-state, homogeneous regimes. *Boundary-layer Meteorology*, 125(2), 167-191.

Zimmerman, J.T.F., 1979. On the Euler-Lagrange transformation and the stokes' drift in the presence of oscillatory and residual currents. *Deep Sea Research Part A. Oceanographic Research Papers*, 26(5), 505-520.

Zuo, S.-h., Zhang, N.-c., Li, B., Chen, S.-l., 2012. A study of suspended sediment concentration in Yangshan deep-water port in Shanghai, China. *International journal of sediment research*, 27(1), 50-60.

List of Figures

List of Tables

About the author

WAN Yuanyang was born on July 09, 1981 in Hubei Province, China. He studied Sediment Transport and River Dynamics in Wuhan University from 1999 to 2003. After that, he obtained an MSc degree on the major of Hydraulics and River Mechanics from the Changjiang River Scientific Research Institute in 2007. After MSc graduation, Wan worked as a researcher on coastal engineering and waterway development in the numerical research department of the Shanghai Estuarine and Coastal Science Research Center (ECSRC). From Nov. 2009 to 2014 Dec., with the help of Mr. Gao and Prof. Dano, Wan traveled to Holland and started a sandwich (part-time) PhD study in UNESCO-IHE. In 2013, Wan was promoted to senior researcher in ECSRC. During his professional experience, Wan was responsible for many important researches and engineering works related to numerical modeling, field observation and laboratory experiment. He published more than 40 peer-reviewed Chinese and English papers. Wan was awarded 4 provincial science and technology prizes in China from 2011 to 2014. He was officially granted 4 Chinese patents on sediment experiment facility. In addition, Wan worked as peer reviewer for various Chinese and international academic journals.

List of Publications

Peer-reviewed international journal articles:

Wan, Y., D. Roelvink, W. Li, D. Qi, and F. Gu (2014), Observation and modeling of the storm-induced fluid mud dynamics in a muddy-estuarine navigational channel, *Geomorphology*, 217(0), 23-36, doi:10.1016/j.geomorph.2014.03.050.

Wan, Y., F. Gu, H. Wu, and D. Roelvink (2014), Hydrodynamic evolutions at the Yangtze Estuary from 1998 to 2009, *Applied Ocean Research*, 47(0), 291-302, doi:10.1016/j.apor.2014.06.009.

Wan, Y., H. Wu, D. Roelvink, and F. Gu (2015), Experimental study on fall velocity of fine sediment in the Yangtze Estuary, China, *Ocean Engineering*, doi:10.1016/j.oceaneng.2015.04.076.

Wan, Y., D. Roelvink, G. Liu, and D. Zhao (2015), Observation of saltwater intrusion and ETM dynamics in a stably stratified estuary: the Yangtze Estuary, China, (preliminary accepted in *Estuarine, Coastal and Shelf Science*).

Wan, Y., L. Wang, and D. Roelvink, Numerical investigation on the factors controlling vertical structure of current, salinity and SSC within a turbidity plume zone, (submitted).

Wan, Y. and D. Roelvink, Modeling of seasonal variations of fine-grained suspended sediment dynamics in the Yangtze Estuary, (in preparation).

Gu, F., L. Mu, D. Qi, J. Li, L. Kong, and Y. Wan (2011), Study on roughness coefficient for unsubmerged reed in the Changjiang Estuary, *Acta Oceanologica Sinica*, 30(5), 108-113, doi:10.1007/s13131-011-0153-0.

Selected peer-reviewed Chinese journal articles(with English abstract):

Wan Y., L. Kong, D. Qi, F. Gu, and W. Wang (2010), Study on characteristics of hydrodynamic and morphological evolution at Hengsha Watercourse of Yangtze Estuary, China, *Journal of Waterway and Harbor*, 31(5), 373-378.

Wan Y., X. Chen, W. Huang, and Q. Shen (2013), Study on the tidalflat currents and navigational channel regulation of the Tongzhou shoal, *Port & waterway Engineering*, 482(0), 8-14.

Wan Y., X. Chen, W. Huang, and Q. Shen (2013), Experimental study on the consolidation velocity of fine sediment in the Yangtze estuary, *Port & waterway Engineering*, 485(0), 111-114.

Wan Y., H. Wu, and F. Gu (2014), Study on settling velocity of fine sediment in the Changjiang River Estuary, *Yangtze River*, 45(1), 98-101.

Wan Y., H. Wu, Q. Shen, and F. Gu (2014), Settling velocity of fine sediment in a tidal environment 1: Definition & study methods, *Port & waterway Engineering*, 489(0), 18-23.

Wan Y., H. Wu, Q. Shen, and F. Gu (2014), Settling velocity of fine sediment in a tidal environment 2: Formulation, *Port & waterway Engineering*, 490(0), 16-20.

Wan Y., H. Wu, Q. Shen, and F. Gu (2014), Settling velocity of fine sediment in a tidal environment 3: Controlling factor, *Port & waterway Engineering*, 491(0), 21-25.

Li W., and Y. Wan (2014), Relationship between back-silting and hydrodynamics in deepwater navigation channel of the Yangtze River estuary, *Hydro-Science and Engineering*, 5(0), 29-33.

Wan Y. (2015), Experimental study on settling velocity of suspended sediments in the Yangtze Estuary, *Marine Sciences*, doi: 10.11759/hykx20130721002.

Selected conference participation:

Wan Y., and D. Qi, Preliminary Analysis on the impact of different stages of Yangtze Estuary Deepwater Channel Regulation Project. Proceedings of the third International Conference on Estuaries and Coasts. Sendai, Japan, Sep. 2009, Volume 1, 181-187.

Wan Y., A New Pattern of Mathematical Model,Preliminary Web-based Platform of Yangtze Estuary circulation model. NCK Days 2010. Westkapelle, The Netherlands. March 2010. Poster.

Kong L., Y. Wan, F. Gu, D. Qi, and W. Wang, Calculation of typhoon-induced sudden siltation using effective wind energy in Yangtze Estuary, 29th ASME International Conference on Ocean, Offshore and Arctic Engineering, 2010/6/6-11, pp 565-568, Shanghai, 2010.

Wan Y., and D. Roelvink, Study and analysis on the variation of coastal environmental gradients at Yangtze Estuary over the past decade. The 11th International Symposium on River Sedimentation, ISRS2010, September 6-9, 2010. Stellenbosch, South Africa. On CD-ROM. Oral Presentation.

Wan Y., Preliminary evaluation the impacts of sea level rise on North Passage Deepwater Navigation Channel of Yangtze Estuary, China. October, 2011. Nanjing, China. The 8th Chinese conference on sediment research. Oral Presentation. (in Chinese)

Wan Y., and D. Roelvink, Monitoring fluid mud in the North Passage navigation channel of Yangtze Estuary, China. the 6th International Conference on Asian and Pacific Coasts. APAC2011. December 14-16, 2011. HongKong, China. On CD-ROM. Oral Presentation.

T - #0664 - 101024 - C0 - 240/170/11 - PB - 9781138028449 - Gloss Lamination